Lecture Notes in Economics and Mathematical Systems

495

Springer-Verlag Berlin Heidelberg GmbH

Igor Konnov

Combined
Relaxation Methods
for Variational Inequalities

 Springer

Author

Prof. Igor Konnov
Department of Applied Mathematics
Kazan University
ul. Kremlevskaya, 18
Kazan 420008, Russia

Cataloging-in-Publication data applied for

Die Deutsche Bibliothek - CIP-Einheitsaufnahme

Konnov, Igor':
Combined relaxation methods for variational inequalities / Igor
Konnov. - Berlin ; Heidelberg ; New York ; Barcelona ; Hong Kong ;
London ; Milan ; Paris ; Singapore ; Tokyo : Springer, 2001
 (Lecture notes in economics and mathematical systems ; 495)
 ISBN 978-3-540-67999-8   ISBN 978-3-642-56886-2 (eBook)
 DOI 10.1007/978-3-642-56886-2

ISSN 0075-8442
ISBN 978-3-540-67999-8

© Springer-Verlag Berlin Heidelberg 2001
Originally published by Springer-Verlag Berlin Heidelberg in 2001

Typesetting: Camera ready by author
Printed on acid-free paper   SPIN: 10781925   42/3142/du   5 4 3 2 1 0

The author dedicates this book to his parents

Preface

Variational inequalities proved to be a very useful and powerful tool for investigation and solution of many equilibrium type problems in Economics, Engineering, Operations Research and Mathematical Physics. In fact, variational inequalities for example provide a unifying framework for the study of such diverse problems as boundary value problems, price equilibrium problems and traffic network equilibrium problems. Besides, they are closely related with many general problems of Nonlinear Analysis, such as fixed point, optimization and complementarity problems. As a result, the theory and solution methods for variational inequalities have been studied extensively, and considerable advances have been made in these areas.

This book is devoted to a new general approach to constructing solution methods for variational inequalities, which was called the *combined relaxation* (CR) approach. This approach is based on combining, modifying and generalizing ideas contained in various relaxation methods. In fact, each combined relaxation method has a two-level structure, i.e., a descent direction and a stepsize at each iteration are computed by finite relaxation procedures. These parameters enable one to determine a hyperplane separating the current iterate and the solution set. Therefore, an iteration sequence of each CR method is Fejer-monotone. Varying the rules of choosing the parameters and auxiliary procedures, one can obtain a number of algorithms within this framework. The author began his investigations in this field in 1991, and published his first work on CR methods for variational inequalities in 1993, see [103]. Since then, a number of CR type methods were designed. It turned out that the CR framework is rather flexible and allows one to construct methods both for single-valued and for multi-valued variational inequalities, including nonlinearly constrained problems. The other essential feature of all the CR methods is that they are convergent under very mild assumptions. In fact, this is the case if there exists a solution to the dual formulation of the variational inequality problem. This assumption is for instance weaker than the pseudomonotonicity of the underlying mapping. Various rates of convergence have been established for most CR methods. In general, they correspond to linear convergence. These properties cause us to anticipate that CR methods might be successfully used for solving diverse applied problems reduced to variational inequalities.

The book can be viewed as an attempt to describe the existing combined relaxation methods as a whole.

In Chapters 1 and 2, we describe various CR methods for variational inequalities with single-valued and multivalued mappings, respectively. The first sections of both the chapters contain the corresponding theoretical background.

Chapters 3 and 4 contain additional results: Chapter 3 is devoted to applications of CR methods. To make the exposition more self-contained, in Section 3.1 we describe other known methods which are applicable for variational inequalities under mild assumptions. We compare their convergence results and numerical performance with those of CR methods. In Section 3.2, we analyse possible applications of CR methods to several economic equilibrium problems. Section 3.3 provides additional results of numerical experiments with test problems.

In Section 4.1, several implementable variants of feasible quasi-nonexpansive mappings, which enable one to apply most CR methods to nonlinearly constrained problems, are described. In Section 4.2, we give new error bounds for linearly constrained variational inequalities. In Section 4.3, the simplest relaxation subgradient method for convex minimization and its rate of convergence are given. This method is a basis for constructing CR methods in the multivalued case. Moreover, it requires no a priori information or line searches, but attains a linear rate of convergence.

The author hopes that the book will draw the additional attention to investigations and applications of CR methods and will contribute the development of new, more powerful solution methods for variational inequalities and related problems.

The author would like to thank all the people who helped him in creating and publishing this book, whose first version appeared on August 24, 1998. Especially, I am grateful to my wife and Mrs. Larissa Gusikhina for their assistance in creating the LaTeX copy of this book. Author's investigations contained in the book were supported in part by RFBR and by the R.T. Academy of Sciences.

Kazan, August 2000   *Igor Konnov*

Table of Contents

Notation and Convention

As usual, we denote by R^n the real n-dimensional Euclidean space, its elements being column vectors. We use superscripts to denote different vectors, the superscript T denotes transpose. Subscripts are used to denote different scalars or components of vectors. For any vectors x and y of R^n, we denote by $\langle x, y \rangle$ their scalar product, i.e.,

$$\langle x, y \rangle = x^T y = \sum_{i=1}^{n} x_i y_i,$$

and by $\|x\|$ the Euclidean norm of x, i.e., $\|x\| = \sqrt{\langle x, x \rangle}$.

Let X be a set in R^n, then int X, $\overline{X}$, and conv$X$ denote the interior, the closure, and the convex hull of X. Also, $\Pi(X)$ denotes the power set of X, i.e., the family of all nonempty subsets of X, and cone X denotes the conic hull of X, i.e.,

$$\text{cone}X = \{q \in R^n \mid q = \lambda t, \lambda \geq 0, t \in X\},$$

and vol (X) denotes the volume of X. We denote by $\mathcal{F}(X)$ the class of feasible quasi-nonexpansive mappings with respect to the set X, i.e., $P \in \mathcal{F}(X)$ if, for every $u \in R^n$, we have $P(u) \in X$ and

$$\|P(u) - x\| \leq \|u - x\| \qquad \forall x \in X.$$

Let x be a point in R^n. Then $d(x, X)$ and $\pi_X(x)$ denote the distance from x to X and the projection of x onto X, respectively:

$$
\begin{aligned}
d(x, X) &= \inf\{\|y - x\| \mid y \in X\}, \\
\pi_X(x) &\in X, \|x - \pi_X(x)\| = d(x, X).
\end{aligned}
$$

We denote by $N(X, x)$ the normal cone to X at x:

$$N(X, x) = \{q \in R^n \mid \langle q, v - x \rangle \leq 0 \quad \forall v \in X\},$$

and by NrX the element of X nearest to origin, i.e., Nr$X = \pi_X(0)$. Also, $B(x, \varepsilon)$ (respectively, $U(x, \varepsilon)$) denotes the closed (respectively, open) Euclidean ball of a radius ε around x:

$$B(x, \varepsilon) \;=\; \{y \in R^n \mid \|y - x\| \leq \varepsilon\},$$
$$U(x, \varepsilon) \;=\; \{y \in R^n \mid \|y - x\| < \varepsilon\}.$$

Besides, $S(x, \varepsilon)$ denotes the Euclidean sphere of a radius ε around x:

$$S(x, \varepsilon) = \{y \in R^n \mid \|y - x\| = \varepsilon\}.$$

If the set X is a cone, we denote by X' its conjugate, i.e.,

$$X' = \{y \in R^n \mid \langle y, x \rangle \geq 0 \quad \forall v \in X\}.$$

For any points x and y of R^n, we set $[x, y] = \operatorname{conv}\{x, y\}$. The segments $(x, y]$, $[x, y)$ and (x, y) are defined analogously. Next,

$$R^n_+ = \{x \in R^n \mid x_i \geq 0 \; i = 1, \ldots, n\}$$

denotes the non- negative orthant in R^n, Z_+ denotes the set of non-negative integers, and I denotes the $n \times n$ unit matrix. For a sequence of numbers $\{\alpha_k\}$, $\liminf_{k \to \infty} \alpha_k$ (respectively, $\limsup_{k \to \infty} \alpha_k$) denotes its lower (respectively, upper) limit. Next, $\phi'(t)$ denotes the derivative of a function $\phi : R \to R$ at a point t, $\nabla F(x)$ denotes either the gradient of a function $F : R^n \to R$ or the Jacobian of a mapping $F : R^n \to R^n$ at x, $\partial F(x)$ denotes the subdifferential of a function $F : R^n \to R$ at x:

$$\partial F(x) = \{g \in R^n \mid \langle g, p \rangle \leq F^\uparrow(x, p)\}$$

where $F^\uparrow(x, p)$ denotes the upper Clarke-Rockafellar derivative for F at x in the direction p:

$$F^\uparrow(x, p) = \sup_{\varepsilon > 0} \lim_{\substack{y \to_F x \\ \alpha \to 0}} \inf_{d \in B(p, \varepsilon)} ((F(y + \alpha d) - F(y))/\alpha),$$

$y \to_F x$ means that both $y \to x$ and $F(y) \to F(x)$. Also, $F'(x, p)$ denotes the usual derivative for F at x in the direction p:

$$F'(x, p) = \lim_{\alpha \to 0} ((F(x + \alpha p) - F(x))/\alpha).$$

Given a set $X \subseteq R^n$ and a function $f : X \to R$, the problem of minimizing f over X is denoted briefly as follows:

$$\min \to \{f(x) \mid x \in X\}.$$

The symbol $\square$ is used to denote the end of proofs.

In this book, all the definitions, theorems, propositions, corollaries, lemmas, and examples are numbered consequently within each section and have the corresponding composite labels. For example, Theorem 1.3.4 denotes the fourth theorem of Section 1.3. Next, all the methods, procedures, figures, formulas and tables are numbered consequently within each chapter. For example, Method 1.2 denotes the second method of Chapter 1.

1. Variational Inequalities with Continuous Mappings

In this chapter, we consider basic schemes of combined relaxation (CR) methods and implementable algorithms for solving variational inequality problems with continuous single-valued mappings under a finite-dimensional space setting.

1.1 Problem Formulation and Basic Facts

In this section, we give some facts from the theory of variational inequality problems and their relations with other problems of Nonlinear Analysis.

1.1.1 Existence and Uniqueness Results

Let U be a nonempty, closed and convex subset of the n-dimensional Euclidean space R^n, $G : U \to R^n$ a continuous mapping. The *variational inequality problem* (VI for short) is the problem of finding a point $u^* \in U$ such that

$$\langle G(u^*), u - u^* \rangle \geq 0 \quad \forall u \in U. \tag{1.1}$$

Most existence results of solutions for VI's are proved by using various fixed-point theorems.

Proposition 1.1.1. *[72, 47] Suppose at least one of the following assumptions holds:*

(a) The set U is bounded;

(b) there exists a nonempty bounded subset W of U such that for every $u \in U \backslash W$ there is $v \in W$ with

$$\langle G(u), u - v \rangle > 0.$$

Then VI (1.1) has a solution.

It is well known that the solution of VI (1.1) is closely related with that of the following problem of finding $u^* \in U$ such that

$$\langle G(u), u - u^* \rangle \geq 0 \quad \forall u \in U. \tag{1.2}$$

Problem (1.2) may be termed as the dual formulation of VI (in short, DVI). We will denote by U^* (respectively, by U^d) the solution set of problem (1.1) (respectively, problem (1.2)). To obtain relationships between U^* and U^d, we need additional monotonicity type properties of G.

Definition 1.1.1. [165, 69] Let W and V be convex sets in R^n, $W \subseteq V$, and let $Q : V \to R^n$ be a mapping. The mapping Q is said to be

(a) *strongly monotone* on W with constant $\tau > 0$ if for each pair of points $u, v \in W$, we have

$$\langle Q(u) - Q(v), u - v \rangle \geq \tau \|u - v\|^2;$$

(b) *strictly monotone* on W if for all distinct $u, v \in W$,

$$\langle Q(u) - Q(v), u - v \rangle > 0;$$

(c) *monotone* on W if for each pair of points $u, v \in W$, we have

$$\langle Q(u) - Q(v), u - v \rangle \geq 0;$$

(d) *pseudomonotone* on W if for each pair of points $u, v \in W$, we have

$$\langle Q(v), u - v \rangle \geq 0 \quad \text{implies} \quad \langle Q(u), u - v \rangle \geq 0;$$

(e) *quasimonotone* on W if for each pair of points $u, v \in W$, we have

$$\langle Q(v), u - v \rangle > 0 \quad \text{implies} \quad \langle Q(u), u - v \rangle \geq 0;$$

(f) *explicitly quasimonotone* on W if it is quasimonotone on W and for all distinct $u, v \in W$, the relation

$$\langle Q(v), u - v \rangle > 0$$

implies

$$\langle Q(z), u - v \rangle > 0 \text{ for some } z \in (0.5(u + v), u).$$

It follows from the definitions that the following implications hold:

$$(a) \implies (b) \implies (c) \implies (d) \implies (e) \text{ and } (f) \implies (e).$$

The reverse assertions are not true in general. Nevertheless, we can state the following.

Lemma 1.1.1. *(i) Each pseudomonotone mapping on U is explicitly quasimonotone on U.*

(ii) If Q is quasimonotone on U and affine, i.e., $Q(u) = Mu + q$, where $q \in R^n$, M is an $n \times n$ matrix, then Q is explicitly quasimonotone on U.

Proof. Fix $u, v \in U$. In case (i), assume for contradiction that Q is pseudomonotone on U, but $\langle Q(v), u - v \rangle > 0$ and, for any $z \in (0.5(u + v), u)$, we have $\langle Q(z), u - v \rangle \leq 0$, hence $\langle Q(z), v - z \rangle \geq 0$. By pseudomonotonicity, we then have $\langle Q(v), v - z \rangle \geq 0$, hence

$$\langle Q(v), u - v \rangle \leq 0,$$

which is a contradiction. Let us now consider case (ii). If $\langle Mv + q, u - v \rangle > 0$, then, by quasimonotonicity, $\langle Mu + q, u - v \rangle \geq 0$. Set $u_\lambda = \lambda v + (1 - \lambda)u$. It follows that

$$\langle Mu_\lambda + q, u - v \rangle = \lambda \langle Mv + q, u - v \rangle + (1 - \lambda)\langle Mu + q, u - v \rangle > 0$$

for all $\lambda \in (0, 1]$ and, hence for some $\lambda \in (0, 0.5)$. $\square$

Under additional assumptions, the stronger result can be stated.

Lemma 1.1.2. *[195, Theorem 5.1] Let Q be affine and quasimonotone on an open convex set X. Then Q is pseudomonotone on X.*

Now we give the relationships between solution sets of VI and DVI in the pseudomonotone case.

Proposition 1.1.2. *[149, 86]*
  (i) The set U^d is convex and closed.
  (ii) It holds that $U^d \subseteq U^*$.
  (iii) If G is pseudomonotone, then $U^* \subseteq U^d$.

Thus, the existence of a solution of problem (1.2) implies that problem (1.1) is also solvable, but the reverse assertion needs generalized monotonicity assumptions. Since the existence of solutions of DVI will play a crucial role in constructing relatively simple solution methods for VI, we consider this question in detail. First we note that Proposition 1.1.2 (iii) does not hold in the (explicitly) quasimonotone case.

Example 1.1.1. Let $U = [-1, 1]$, $G(u) = u^2$. Then G is explicitly quasimonotone on U and $U^* = \{0\} \bigcup \{-1\}$. However, $U^d = \{-1\}$, since, for example, $G(-1)(-1 - 0) = -1 < 0$.

Moreover, problem (1.2) may even have no solutions in the quasimonotone case.

Example 1.1.2. Let $n = 2$,

$$U = \text{conv}\{u^0, u^1, u^2\},$$

where $u^0 = (-1, 0)^T$, $u^1 = (0, \sqrt{3})^T$, $u^2 = (1, 0)^T$, i.e. U is a equilateral triangle (see Figure 1.1). Choose $\alpha > 0$ and $\varepsilon > 0$ small enough and divide U into the four subsets:

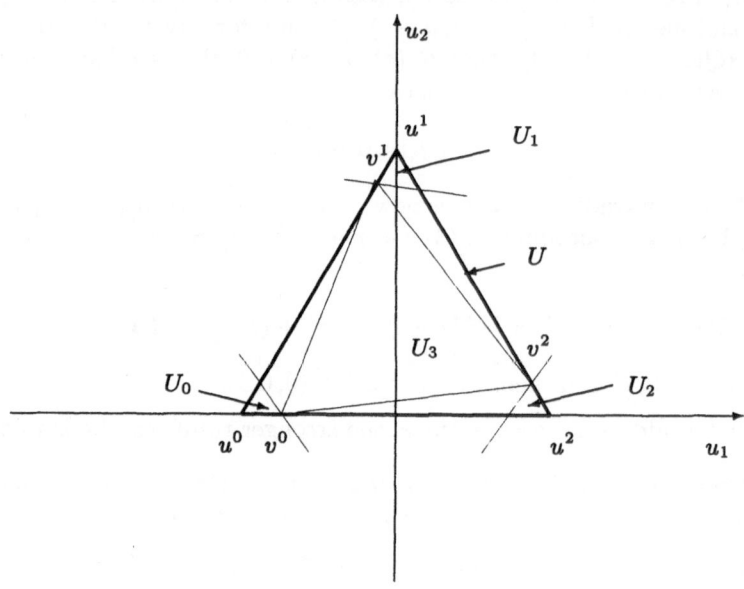

Fig.1.1.

$$U_0 = \left\{ u \in U \mid \sqrt{3}u_1 + u_2 + \alpha\sqrt{3} \leq 0 \right\},$$

$$U_1 = \left\{ u \in U \mid -\varepsilon\sqrt{3}u_1 - u_2 + \alpha\sqrt{3} \leq 0 \right\},$$

$$U_2 = \left\{ u \in U \mid -\sqrt{3}u_1 + u_2 + \alpha\sqrt{3} \leq 0 \right\},$$

$$U_3 = U \setminus \left(\bigcup_{i=0}^{2} U_i \right).$$

We now define the mapping $G : U \to R^2$ as follows:

$$G(u) = \begin{cases} (\sqrt{3}u_1 + u_2 + \alpha\sqrt{3})^2 g^0 & \text{if } u \in U_0, \\ (-\varepsilon\sqrt{3}u_1 - u_2 + \alpha\sqrt{3})^2 g^1 & \text{if } u \in U_1, \\ (-\sqrt{3}u_1 + u_2 + \alpha\sqrt{3})^2 g^2 & \text{if } u \in U_2, \\ 0 & \text{if } u \in U_3; \end{cases}$$

where $g^0 = 0.5(-1, \sqrt{3})^T$, $g^1 = 0.5(1, -\varepsilon\sqrt{3})^T$, $g^2 = -0.5(1, \sqrt{3})^T$. Note that $G(u) = 0$ if $u \in \overline{U_3} \cap U_i, i = 0, 1, 2$ and that G is continuous on U. If we now set

$$F(u) = \{v \in U \mid \langle G(u), u - v \rangle \geq 0\},$$

then it is easy to see that $\bigcap_{i=0}^{2} F(u^i) = \emptyset$. Therefore, $U^d = \emptyset$, but $U^* \neq \emptyset$. We now show that, under an appropriate choice of α and ε, G is quasimonotone on

U. It is clear that G is quasimonotone on each union $U_3 \bigcup U_i, i = 0, 1, 2$. Take $\alpha = 0.9$, $\varepsilon = 0.1$. Define the points $v^0 = (-0.9, 0)^T$, $v^1 = (-1/11, 10\sqrt{3}/11)^T$, $v^2 = (0.95, 0.05\sqrt{3})^T$. Note that if we take every pair of points u, v such that either $u \in U_0, v \in U_1$, or $u \in U_1, v \in U_2$, or $u \in U_2, v \in U_0$, then

$$\langle G(v), u - v \rangle \le 0.$$

On the other hand, if we choose $u \in U_1, v \in U_0$, or $u \in U_2, v \in U_1$, or $u \in U_0, v \in U_2$, then

$$\langle G(v), u - v \rangle \ge 0$$

and the strict inequality holds if $v \notin \overline{U_3} \bigcap U_i, i = 0, 1, 2$. In this case we must have $\langle G(u), u - v \rangle \ge 0$ since

$$
\begin{aligned}
\langle g^1, v^1 - v^0 \rangle &= 59/220 > 0, \\
\langle g^2, v^2 - v^1 \rangle &= 179/220 > 0, \\
\langle g^0, v^0 - v^2 \rangle &= 0.85 > 0
\end{aligned}
$$

(see Figure 1.1). Thus, G is quasimonotone on U, as required.

However, we can give an example of solvable DVI (1.2) with the underlying mapping G which is not quasimonotone.

Example 1.1.3. Let the mapping G be linear, i.e. $G(u) = Mu$ where M is an $n \times n$ matrix. For simplicity, let $n = 2$,

$$U = \{u \in R^2 \mid |u_1 - u_2| + 0.1|u_1 + u_2 - 4| \le 0.2\}.$$

Note that U is convex and closed. If we now choose

$$M = \begin{pmatrix} -0.2 & 10 \\ 10 & 0.5 \end{pmatrix},$$

then $U^* = U^d = \{(1, 1)^T\}$. But the matrix M is not positive semidefinite, hence the mapping G cannot be monotone. Next, the points $u = (1.9, 2.1)^T$ and $v = (2.1, 1.9)^T$ are feasible, moreover, $\langle Mv, u - v \rangle = 0.674 > 0$. But $\langle Mu, u - v \rangle = -0.114 < 0$, hence the mapping G cannot be quasimonotone either.

Thus, the class of quasimonotone mappings and the class of mappings that guarantee for solutions of problem (1.2) to exist have nonempty intersection, but they do not contain each other.

Now we give the conditions under which there exists a solution to problem (1.2) in the explicitly quasimonotone case.

Proposition 1.1.3. *[121, 40] Suppose that $G : U \to R^n$ is explicitly quasi-monotone and that there exist a bounded subset W of U and a point $v \in W$ such that, for every $u \in U \backslash W$, we have*

$$\langle G(v), u - v \rangle > 0.$$

Then $U^d \ne \emptyset$.

In general, VI can have more than one solution. We now recall the conditions under which VI (1.1) has a unique solution.

Proposition 1.1.4. *(e.g. [70])*
   (i) If G is strictly monotone, then VI (1.1) has at most one solution.
   (ii) If G is strongly monotone, then VI (1.1) has a unique solution.

In addition, we now recall the known monotonicity criteria for continuously differentiable mappings.

Proposition 1.1.5. *[165, Theorem 5.4.3] Let W be an open convex subset of V and let $Q : V \to R^n$ be continuously differentiable on W.*
   (i) Q is monotone on W if and only if ∇Q is positive semidefinite on W;
   (ii) Q is strictly monotone on W if ∇Q is positive definite on W;
   (iii) Q is strongly monotone on W with constant τ if and only if it holds that

$$\langle \nabla Q(u)p, p \rangle \geq \tau \|p\|^2 \qquad \forall p \in R^n, u \in W.$$

Note that the Jacobian of a differentiable strictly monotone mapping need not be positive definite; e.g., see [218].

1.1.2 Variational Inequalities and Related Problems

VI's are closely related with many general problems of Nonlinear Analysis, such as complementarity, fixed point and optimization problems. The simplest example of VI is the problem of solving a system of equations. It is easy to see that if $U = R^n$ in (1.1), then VI (1.1) is equivalent to the problem of finding a point $u^* \in R^n$ such that

$$G(u^*) = 0.$$

If the mapping G is affine, i.e., $G(u) = Mu + q$, then the above problem is equivalent to the classical system of linear equations

$$Mu^* = -q. \tag{1.3}$$

Let U be a convex cone in R^n. The *complementarity problem* (CP for short) is to find a point $u^* \in U$ such that

$$G(u^*) \in U', \quad \langle G(u^*), u^* \rangle = 0, \tag{1.4}$$

where U' is the dual cone to U, i.e.,

$$U' = \{v \in R^n \mid \langle u, v \rangle \geq 0 \quad \forall u \in U\}.$$

This problem can be viewed as a particular case of VI, as stated below.

Proposition 1.1.6. *[85] Let U be a convex cone. Then problem (1.1) is equivalent to problem (1.4).*

Among various classes of CP's, the following ones are most investigated. The *standard complementarity problem* corresponds to the case where $U = R^n_+$ in (1.4) and the *linear complementarity problem* (LCP for short) corresponds to the case where $U = R^n_+$ and G is affine, i.e., $G(u) = Mu + q$ in (1.4).

Next, let U be again an arbitrary convex closed set in R^n and let T be a continuous mapping from U into itself. The *fixed point problem* is to find a point $u^* \in U$ such that

$$u^* = T(u^*). \tag{1.5}$$

This problem can be also converted into a VI format.

Proposition 1.1.7. *(e.g., see [89, Section 3.1]) If the mapping G is defined by*

$$G(u) = u - T(u), \tag{1.6}$$

then problem (1.1) coincides with problem (1.5).

Moreover, we can obtain the condition which guarantees the monotonicity of G.

Definition 1.1.2. (e.g. [89, Chapter 1]) A mapping $Q : R^n \to R^n$ is said to be *non-expansive* on V if for each pair of points $u, v \in V$, we have

$$\|Q(u) - Q(v)\| \leq \|u - v\|.$$

Proposition 1.1.8. *If $T : U \to U$ is non-expansive, then the mapping G, defined in (1.6), is monotone.*

Proof. Take any $u, v \in U$. Then we have

$$
\begin{aligned}
\langle G(u) - G(v), u - v \rangle &= \|u - v\|^2 - \langle T(u) - T(v), u - v \rangle \\
&\geq \|u - v\|^2 - \|T(u) - T(v)\|\|u - v\| \geq 0,
\end{aligned}
$$

i.e., G is monotone.   □

Now, we consider the well-known optimization problem. Let $f : U \to R$ be a real-valued function. Then we can define the following *optimization problem* of finding a point $u^* \in U$ such that

$$f(u^*) \leq f(u) \quad \forall u \in U,$$

or briefly,

$$\min \to \{f(u) \mid u \in U\}. \tag{1.7}$$

We denote by U_f the solution set of this problem.

Recall the definitions of convexity type properties for functions.

Definition 1.1.3. [147, 8] Let W and V be convex sets in R^n such that $W \subseteq V$, and let $\varphi : V \to R$ be a differentiable function. The function φ is said to be

(a) *strongly convex* on W with constant $\tau > 0$ if for each pair of points $u, v \in W$ and for all $\alpha \in [0, 1]$, we have

$$\varphi(\alpha u + (1 - \alpha)v) \le \alpha \varphi(u) + (1 - \alpha)\varphi(v) - 0.5\alpha(1 - \alpha)\tau \|u - v\|^2;$$

(b) *strictly convex* on W if for all distinct $u, v \in W$ and for all $\alpha \in (0, 1)$,

$$\varphi(\alpha u + (1 - \alpha)v) < \alpha \varphi(u) + (1 - \alpha)\varphi(v);$$

(c) *convex* on W if for each pair of points $u, v \in W$ and for all $\alpha \in [0, 1]$, we have

$$\varphi(\alpha u + (1 - \alpha)v) \le \alpha \varphi(u) + (1 - \alpha)\varphi(v);$$

(d) *pseudoconvex* on W if for each pair of points $u, v \in W$ and for all $\alpha \in [0, 1]$, we have

$$\langle \nabla\varphi(v), u - v \rangle \ge 0 \quad \text{implies } \varphi(u) \ge \varphi(v);$$

(e) *quasiconvex* on W if for each pair of points $u, v \in W$ and for all $\alpha \in [0, 1]$, we have

$$\varphi(\alpha u + (1 - \alpha)v) \le \max\{\varphi(u), \varphi(v)\};$$

(f) *explicitly quasiconvex* on W if it is quasiconvex on W and for all distinct $u, v \in W$ and for all $\alpha \in (0, 1)$, we have

$$\varphi(\alpha u + (1 - \alpha)v) < \max\{\varphi(u), \varphi(v)\}.$$

The function $\varphi : V \to R$ is said to be *strongly concave* with constant τ (respectively, *strictly concave, pseudoconcave, quasiconcave, explicitly quasiconcave*) on W, if the function $-\varphi$ is strongly convex with constant τ (respectively, strictly convex, convex, pseudoconvex, quasiconvex, explicitly quasiconvex) on W.

It follows directly from the definitions that the following implications hold:

$$(a) \implies (b) \implies (c) \implies (f) \implies (e).$$

The reverse assertions are not true in general. We can also include the pseudoconvexity in the above relations. First recall the well-known property of the gradient of a convex function.

Lemma 1.1.3. *(e.g. [225, Chapter 2, Theorem 3]) Let V be a convex set in R^n. A differentiable function $f : V \to R$ is convex if and only if*

$$f(u) - f(v) \ge \langle \nabla f(v), u - v \rangle \quad \forall u, v \in V. \tag{1.8}$$

From Lemma 1.1.3 it follows the implication $(c) \implies (d)$. Besides, the implication $(d) \implies (f)$ is also true; e.g., see [147, p.113]. We now state the relationships between (generalized) convexity of functions and (generalized) monotonicity of their gradients.

Proposition 1.1.9. *[8, 165, 87, 194, 69] Let W be an open convex subset of V. A differentiable function $f : V \to R$ is strongly convex with constant τ (respectively, strictly convex, convex, pseudoconvex, quasiconvex, explicitly quasiconvex) on W, if and only if its gradient map $\nabla f : U \to R^n$ is strongly monotone with constant τ (respectively, strictly monotone, monotone, pseudomonotone, quasimonotone, explicitly monotone) on W.*

We now give the well-known optimality condition for problem (1.7).

Theorem 1.1.1. *Suppose that $f : U \to R$ is a differentiable function. Then:*
(i) $U_f \subseteq U^*$, i.e., each solution of (1.7) is a solution of VI (1.1), where

$$G(u) = \nabla f(u); \qquad (1.9)$$

(ii) if f is pseudoconvex and G is defined by (1.9), then $U^* \subseteq U_f$.

Proof. Part (ii) is trivial. In case (i), assume, for contradiction, that there exists $u^* \in U_f \backslash U^*$, i.e., there is a point $v \in U$ such that

$$\langle \nabla f(u^*), v - u^* \rangle < 0.$$

Then, for $\alpha > 0$ small enough, we must have $v_\alpha = u^* + \alpha(v - u^*) = \alpha v + (1 - \alpha)u^* \in U$ and

$$f(v_\alpha) = f(u^*) + \alpha \langle \nabla f(u^*), v - u^* \rangle + o(\alpha) < f(u^*),$$

i.e., $u^* \notin U_f$, which is a contradiction. □

Thus, optimization problem (1.7) can be reduced to VI (1.1) with the (generalized) monotone underlying mapping G if the function f in (1.7) possesses the corresponding (generalized) convexity property. However, VI which expresses the optimality condition in optimization enjoys additional properties in comparison with the usual VI. For instance, if f is twice continuously differentiable, then its Hessian matrix $\nabla^2 f = \nabla G$ is symmetric. Conversely, if the mapping $\nabla G : R^n \to R^n \times R^n$ is symmetric, then for any fixed v there exists the function

$$f(u) = \int_0^1 \langle G(v + \tau(u - v)), u - v \rangle d\tau$$

such that (1.9) holds (see [165, Theorem 4.1.6]). It is obvious that the Jacobian ∇G in (1.1) is in general asymmetric. Next, consider the case of convex optimization problem (1.7). In other words, let the function f be convex and differentiable. Then, according to Theorem 1.1.1, (1.7) is equivalent to (1.1) with G being defined in (1.9). Due to Proposition 1.1.9, the mapping ∇f is monotone. Besides, due to Lemma 1.1.3 we see that for each $u \in U \backslash U^*$, we have

$$\langle \nabla f(u), u - u^* \rangle > 0; \tag{1.10}$$

i.e., $-\nabla f(u)$ makes an acute angle with any vector $u^* - u$ at each non optimal point $u \in U$. In the general case this property does not hold. Namely, consider problem (1.1) with G being monotone. Then, at each $u \in U \backslash U^*$, we only have

$$\langle G(u), u - u^* \rangle \geq 0$$

due to Proposition 1.1.2, i.e., the angle between $-G(u)$ and $u - u^*$ need not be acute. We now give the classical example of such a problem.

Example 1.1.4. Let $U = R^2$, $G(u) = (u_2, -u_1)^T$. Then G is monotone, $U^* = \{(0,0)^T\}$, but for any $u \notin U^*$ we have

$$\langle G(u), u - u^* \rangle = u_2 u_1 - u_1 u_2 = 0,$$

i.e., the property similar to (1.10) does not hold. It should be also noted that the Jacobian

$$\nabla G(u) = \begin{pmatrix} 0 & 1 \\ -1 & 0 \end{pmatrix}$$

is asymmetric, so that there is no function f satisfying (1.9).

On the other hand, combining Theorem 1.1.1 and Propositions 1.1.1 and 1.1.4 yields existence and uniqueness results for optimization problems.

Corollary 1.1.1. *Let W be an open convex set in R^n such that $U \subseteq W$. Let $f : W \rightarrow R$ be a differentiable and strongly convex function. Then there exists a unique solution to problem (1.7).*

Proof. Indeed, from Theorem 1.1.1 it follows that problem (1.7) is equivalent to (1.1), (1.9), moreover, $G = \nabla f$ is strongly monotone due to Proposition 1.1.9. The desired result now follows from Proposition 1.1.4 (ii). $\qquad \square$

Saddle point problems are closely related to optimization as well as to noncooperative game problems. Let X be a convex closed set in R^l and let Y be a convex closed set in R^m. Suppose that $L : R^l \times R^m \rightarrow R$ is a differentiable convex-concave function, i.e., $L(\cdot, y)$ is convex for each $y \in Y$ and $L(x, \cdot)$ is concave for each $x \in X$. The *saddle point problem* is to find a pair of points $x^* \in X$, $y^* \in Y$ such that

$$L(x^*, y) \leq L(x^*, y^*) \leq L(x, y^*) \quad \forall x \in X, \forall y \in Y. \tag{1.11}$$

Set $n = l + m$, $U = X \times Y$ and define the mapping $G : R^n \rightarrow R^n$ as follows:

$$G(u) = G(x, y) = \begin{pmatrix} \nabla_x L(x, y) \\ -\nabla_y L(x, y) \end{pmatrix}. \tag{1.12}$$

From Theorem 1.1.1 we now obtain the following equivalence result.

Corollary 1.1.2. *Problems (1.11) and (1.1), (1.12) are equivalent.*

It should be noted that G in (1.12) is also monotone, see [185, Section 37].

Saddle point problems are proved to be a useful tool for "eliminating" functional constraints in optimization. Let us consider the optimization problem

$$\min \rightarrow \{f_0(x) \mid x \in D\}, \tag{1.13}$$

where

$$D = \{x \in X \mid f_i(x) \le 0 \quad i = 1, \dots, m\}, \tag{1.14}$$

$f_i : R^l \rightarrow R$, $i = 0, \dots, m$ are convex differentiable functions,

$$X = \{x \in R^l \mid x_j \ge 0 \quad \forall j \in J\}, J \subseteq \{1, \dots, l\}. \tag{1.15}$$

Then we can define the Lagrange function associate to problem (1.13) – (1.15) as follows:

$$L(x, y) = f_0(x) + \sum_{i=1}^{m} y_i f_i(x). \tag{1.16}$$

To obtain the relationships between problems (1.13) – (1.15) and (1.11), (1.16), we need certain constraint qualification conditions. Namely, consider the following assumption.

(C) *Either all the functions f_i, $i = 1, \dots, m$ are affine, or there exists a point $\bar{x}$ such that $f_i(\bar{x}) < 0$ for all $i = 1, \dots, m$.*

Proposition 1.1.10. *(e.g. [5, Chapters 3–5], [214, Chapter 4])*
(i) If (x^*, y^*) is a saddle point of the function L in (1.16) with $Y = R_+^m$, then x^* is a solution to problem (1.13) – (1.15).
(ii) If x^* is a solution to problem (1.13) – (1.15) and condition (C) holds, then there exists a point $y^* \in Y = R_+^m$ such that (x^*, y^*) is a solution to the saddle point problem (1.11), (1.16).

By using Corollary 1.1.2 we now see that optimization problem (1.13) – (1.15) can be replaced by VI (1.1) (or equivalently, by CP (1.4) since X is a convex cone), where $U = X \times Y$, $Y = R_+^m$, $f(x) = (f_1(x), \dots, f_m(x))^T$, and

$$G(u) = \begin{pmatrix} \nabla f_0(x) + \sum_{i=1}^{m} y_i \nabla f_i(x) \\ -f(x) \end{pmatrix} \tag{1.17}$$

with G being monotone.

Similarly, we can convert VI with functional constraints into VI (or CP) with simple constraints. Let us consider the following problem of finding $x^* \in D$ such that

$$\langle F(x^*), x - x^* \rangle \geq 0 \quad \forall x \in D, \tag{1.18}$$

where $F : R^l \to R^l$ is a continuous mapping, D is the same as in (1.14), (1.15).

Proposition 1.1.11. *(i) If $u^* = (x^*, y^*)$ is a solution to (1.1) with*

$$U = X \times Y, Y = R_+^m, G(u) = G(x, y) = \begin{pmatrix} F(x) + \sum_{i=1}^m y_i \nabla f_i(x) \\ -f(x) \end{pmatrix}, \tag{1.19}$$

then x^* is a solution to problem (1.18), (1.14), (1.15).

(ii) If condition (C) holds and x^* is a solution to problem (1.18), (1.14), (1.15), then there exists a point $y^* \in Y = R_+^m$ such that (x^*, y^*) is a solution to (1.1), (1.19).

Proof. (i) Let (x^*, y^*) be a solution to (1.1), (1.19), or equivalently, to the following system:

$$\langle F(x^*) + \sum_{i=1}^m y_i^* \nabla f_i(x^*), x - x^* \rangle \geq 0 \quad \forall x \in X,$$

$$\langle -f(x^*), y - y^* \rangle \geq 0 \quad \forall y \in Y = R_+^m. \tag{1.20}$$

The last relation implies $x^* \in D$ and $\langle f(x^*), y^* \rangle = 0$, so that, applying the first relation in (1.20) together with Lemma 1.1.3, we have

$$0 \leq \langle F(x^*), x - x^* \rangle + \sum_{i=1}^m y_i^* (f_i(x) - f_i(x^*)) \leq \langle F(x^*), x - x^* \rangle \quad \forall x \in D,$$

i.e., x^* is a solution to problem (1.18), (1.14), (1.15).

(ii) If x^* is a solution to problem (1.18), (1.14), (1.15), it is a solution to the following convex optimization problem

$$\min \to \{ \langle F(x^*), x \rangle \mid x \in D \},$$

and due to Proposition 1.1.10 (ii) there exists a point $y^* \in Y = R_+^m$ such that (x^*, y^*) is a saddle point of the function $\tilde{L}(x, y) = \langle F(x^*), x \rangle + \sum_{i=1}^m y_i f_i(x)$. Now the left inequality in (1.11) implies the second inequality in (1.20), whereas the right inequality in (1.11) implies that x^* is a solution to the following convex optimization problem:

$$\min \to \{ \langle F(x^*), x \rangle + \sum_{i=1}^m y_i^* f_i(x) \mid x \in X \},$$

which is equivalent to the first relation in (1.20) due to Theorem 1.1.1. Therefore, (x^*, y^*) is a solution to (1.1), (1.19). □

1.2 Main Idea of CR Methods

In this section, we describe the basic algorithmic scheme for combined relaxation (CR) methods and the general approach to compute iteration parameters in such methods.

1.2.1 Newton-like Iterations

One of most popular approaches for solving general problems of Nonlinear Analysis consists of creating a sequence $\{u^k\}$ such that each u^{k+1} is a solution of some auxiliary problem, which can be viewed as an approximation of the initial problem at the previous point u^k. For instance, if the underlying mapping (function) is differentiable, it can be replaced by its linear approximation.

At the beginning we consider the simplest problem of solving the nonlinear equation

$$\phi(t) = 0, \tag{1.21}$$

where $\phi : R \to R$ is continuously differentiable. Recall that the Newton method being applied to this problem iteratively solves the linearized problem

$$\phi(t_k) + \phi'(t_k)(t - t_k) = 0, \tag{1.22}$$

where t_k is a current iteration point. Obviously, we obtain the well-known process

$$t_{k+1} := t_k - \phi(t_k)/\phi'(t_k), \tag{1.23}$$

which, under certain assumptions, converges quadratically to a solution of (1.21).

There exist numerous modifications and extensions of the Newton method. Obviously, the linearization approach can be implemented in the case of VI (1.1), where G is continuously differentiable. Namely, the Newton method being applied to VI (1.1) consists of solving the linearized problem of finding a point $\bar{z} \in U$ such that

$$\langle G(u^k) + \nabla G(u^k)(\bar{z} - u^k), v - \bar{z} \rangle \geq 0 \quad \forall v \in U, \tag{1.24}$$

where u^k is a current iteration point and setting $u^{k+1} := \bar{z}$ for $k = 0, 1, \dots$ It is clear that problem (1.24) is a direct extension of (1.22). It can be also shown that, under certain assumptions, the Newton method converges quadratically to a solution of VI (1.1) [82, 170]. However, the evaluation of the Jacobian matrix $\nabla G(u^k)$ at each iteration can be a too hard problem, besides, the Jacobian can be non positive definite. It is also known that the Newton method has only local convergence properties. Most modifications of the Newton method consist of replacing the Jacobian $\nabla G(u^k)$ in (1.24) with an $n \times n$ positive definite matrix A_k. One can then obtain various Newton–like methods such as quasi–Newton methods, successive overrelaxation methods, etc. We now consider the further extension of the Newton method which follows the general iterative schemes of [36, 32].

Definition 1.2.1. Let V be a convex set in R^n and let $T : V \to R^n$ be a mapping. The mapping T is said to be *Lipschitz continuous* with constant L if for each pair of points $u, v \in V$, we have

$$\|T(u) - T(v)\| \le L\|u - v\|.$$

The mapping T is said to be *locally Lipschitz continuous* if it is Lipschitz continuous on each bounded subset of V.

Note that the class of locally Lipschitz continuous mappings on V strictly contains the class of Lipschitz continuous mappings on V unless V is bounded.

Let V be a convex set in R^n such that $U \subseteq V$. Let us define a family of mappings $\{T_k : V \times V \to R^n\}$ such that, for each $k = 0, 1, \ldots$,

(A1) $T_k(u, \cdot)$ *is strongly monotone with constant* $\tau_k' > 0$ *for every* $u \in U$;

(A2) $T_k(u, \cdot)$ *is Lipschitz continuous with constant* $\tau_k'' > 0$ *for every* $u \in V$;

(A3) $T_k(u, u) = 0$ *for every* $u \in V$.

Denote by $\nabla_z T_k(u, \cdot)$ the Jacobian of the mapping $T_k(u, \cdot)$. Now, consider the iterative method in which problem (1.24) is replaced by the problem of finding a point $\bar{z} \in U$ such that

$$\langle G(u^k) + \lambda_k^{-1} T_k(u^k, \bar{z}), v - \bar{z} \rangle \ge 0 \quad \forall v \in U. \tag{1.25}$$

We first establish several properties of solutions for problem (1.25).

Lemma 1.2.1. (i) *Problem (1.25) has a unique solution.*

(ii) *It holds that*

$$\langle G(u^k), u^k - \bar{z} \rangle \ge \lambda_k^{-1} \langle T_k(u^k, \bar{z}), \bar{z} - u^k \rangle \ge \lambda_k^{-1} \tau_k' \|\bar{z} - u^k\|^2. \tag{1.26}$$

(iii) *The solution $\bar{z}$ of (1.25) coincides with $u^k \in U$ if and only if $u^k \in U^*$.*

Proof. Assertion (i) follows directly from Property (A1) and from Proposition 1.1.4 (ii). Next, since $\bar{z}$ is a solution of (1.25), we have

$$\langle G(u^k), u^k - \bar{z} \rangle + \lambda_k^{-1} \langle T_k(u^k, \bar{z}), u^k - \bar{z} \rangle \ge 0,$$

i.e., the first inequality in (1.26) holds. By the properties of $T(u^k, \cdot)$, we obtain

$$\langle T_k(u^k, \bar{z}), \bar{z} - u^k \rangle = \langle T_k(u^k, \bar{z}) - T_k(u^k, u^k), \bar{z} - u^k \rangle \ge \tau_k' \|\bar{z} - u^k\|^2,$$

hence, the second inequality in (1.26) holds, too. Next, if $\bar{z} = u^k$, then from (1.25) and Property (A3) we have

$$\langle G(u^k), v - u^k \rangle = \langle G(u^k) + \lambda_k^{-1} T_k(u^k, u^k), v - u^k \rangle \geq 0 \quad \forall v \in U,$$

i.e., $u^k \in U^*$. Now, suppose that $u^k \in U^*$ and $\bar{z} \neq u^k$. Then, by (1.26),

$$\langle G(u^k), \bar{z} - u^k \rangle \leq -\lambda_k^{-1} \tau_k' \|\bar{z} - u^k\|^2 < 0,$$

so that $u^k \notin U^*$. By contradiction, we see that assertion (iii) is also true. $\square$

Thus, the method above is well defined and gives an optimality condition for the source problem (1.1). Nevertheless, it is desirable that there exist an effective algorithm for solving problem (1.25). For instance, taking into account Proposition 1.1.5, we can choose

$$T_k(u, z) = A_k(z - u) \tag{1.27}$$

where A_k is an $n \times n$ positive definite matrix. Then problem (1.25) becomes much simpler than the initial VI (1.1). Indeed, it coincides with a system of linear equations when $U = R^n$ or with a linear complementarity problem (LCP) when $U = R_+^n$. Besides, due to Proposition 1.1.11, problem (1.25) reduces to LCP in the case where the feasible set U is defined by affine constraints. It is well known that such problems can be solved by finite algorithms.

On the other hand, in order to maintain fast convergence it is desirable that A_k (or $\nabla_z T_k(u^k, \bar{z})$) approximate $\nabla G(u^k)$. Obviously, if $\nabla G(u^k)$ is positive definite, we can simply choose $A_k = \nabla G(u^k)$. Then problem (1.25), (1.27) coincides with (1.24). Moreover, we can follow the Levenberg–Marquardt approach, i.e., set

$$A_k = \nabla G(u^k) + \beta_k I$$

or make use of an appropriate quasi-Newton update. These approaches are applicable even if $\nabla G(u^k)$ is not positive definite.

Note that the matrix $\nabla G(u^k)$ is usually asymmetric and the approximate matrix A_k can maintain this property. However, if we choose A_k in (1.27) to be symmetric, then the solution of problem (1.25) coincides with the projection of the point $u^k - A_k^{-1} G(u^k)$ onto U in the norm scaled by A_k, as the following lemma states.

Lemma 1.2.2. *Suppose that A_k in (1.27) is symmetric and positive definite. A point $\bar{z} \in U$ is a solution of (1.25), (1.27), if and only if it solves the problem:*

$$\min \rightarrow \{\psi_k(z) \mid z \in U\}, \tag{1.28}$$

where

$$\psi_k(z) = \langle G(u^k), z - u^k \rangle + 0.5\lambda_k^{-1} \langle A_k(z - u^k), z - u^k \rangle, \tag{1.29}$$

or equivalently,

$$\psi_k(z) = \langle A_k(z - [u^k - \lambda_k A_k^{-1} G(u^k)]), z - [u^k - \lambda_k A_k^{-1} G(u^k)] \rangle. \tag{1.30}$$

Proof. Since ψ_k in (1.29) is strongly convex, problem (1.28), (1.29) must have a unique solution. Next, due to Theorem 1.1.1, the necessary and sufficient condition of optimality for problem (1.28), (1.29) is the following:

$$\langle \nabla \psi_k(\bar{z}), v - \bar{z} \rangle \geq 0 \quad \forall v \in U,$$

which coincides with (1.25), (1.27). On the other hand, we see that

$$
0.5\lambda_k^{-1}\langle A_k(z - [u^k - \lambda_k A_k^{-1}G(u^k)]), z - [u^k - \lambda_k A_k^{-1}G(u^k)]\rangle
$$
$$
= \quad 0.5\lambda_k^{-1}\langle A_k(z - u^k), z - u^k\rangle + 0.5\langle G(u^k), z - u^k\rangle
$$
$$
+ \quad 0.5\langle A_k(z - u^k), A_k^{-1}G(u^k)\rangle + 0.5\lambda_k\|A_k^{-1}G(u^k)\|^2
$$
$$
= \quad \langle G(u^k), z - u^k\rangle + 0.5\lambda_k^{-1}\langle A_k(z - u^k), z - u^k\rangle + 0.5\lambda_k\|A_k^{-1}G(u^k)\|^2,
$$

i.e., to obtain (1.29) we need to multiply the expression in (1.30) by $0.5\lambda_k^{-1} > 0$ and subtract the constant term $0.5\lambda_k\|A_k^{-1}G(u^k)\|^2$. Obviously, the solution of both the problems (1.28), (1.29) and (1.28), (1.30) must be unique. □

Therefore, the simplest choice $A_k \equiv I$ in (1.27) leads to the well-known projection method (see [202]), i.e., in this case the method (1.25), (1.27) can be written as follows

$$u^{k+1} := \pi_U(u^k - \lambda_k G(u^k)). \tag{1.31}$$

It follows that, in the symmetric case, the solution $\bar{z}$ of the auxiliary problem (1.25) can be found by rather simple algorithms. Thus, the described Newton-like scheme in fact contains a class of iterative solution methods.

However, all these methods require restrictive assumptions either G be strictly monotone or its Jacobian be symmetric for convergence, as the following example illustrates.

Example 1.2.1. Let U and G be the same as those in Example 1.1.4, i.e., $U = R^2, G(u) = (u_2, -u_1)^T$. As indicated in Example 1.1.4, G is monotone and $U^* = (0,0)^T$. For any $\lambda_k > 0$, we have

$$\|u^k - \lambda_k G(u^k)\|^2 = (u_1^k - \lambda_k u_2^k)^2 + (u_2^k + \lambda_k u_1^k)^2 = (1 + \lambda_k^2)\|u^k\|^2 > \|u^k\|^2.$$

Thus, method (1.31) fails to provide convergence regardless the stepsize rule (see Figure 1.2). Note that this conclusion is in general true for all the Newton-like methods and it is caused by the fact that the angle between $G(u^k)$ and $u^k - u^*$ with $u^* \in U^*$ need not be acute. On the other hand, if G is a monotone gradient mapping or a strictly monotone mapping, then the angle is acute, see (1.10).

Therefore, most well-known differentiable optimization methods cannot be extended directly for the case of VI. In the next section, we describe a general approach to constructing iterative solution methods, which involves a solution of the auxiliary problem (1.25) in order to compute iteration parameters in the main process. As a result, we prove convergence of such a process under rather mild assumptions.

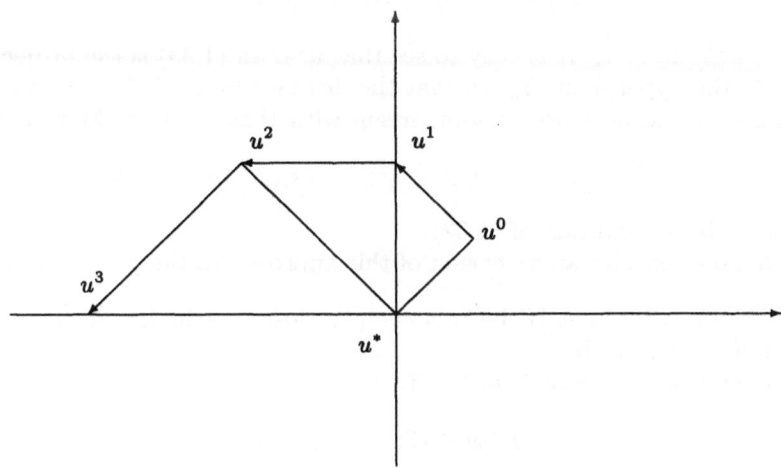

Fig.1.2.

1.2.2 Basic Properties of CR Methods

We first consider another approach to extend the Newton method (1.23).
Suppose $f : R^n \rightarrow R$ is a non-negative, continuously differentiable and convex
function. Let us consider the problem of finding a point $u^* \in R^n$ such that

$$f(u^*) = 0. \tag{1.32}$$

Its solution can be found by the following gradient process

$$u^{k+1} := u^k - (f(u^k)/\|\nabla f(u^k)\|^2)\nabla f(u^k). \tag{1.33}$$

Note that process (1.33) can be viewed as some extension of (1.23). Indeed,
the next iterate u^{k+1} also solves the linearized problem

$$f(u^k) + \langle \nabla f(u^k), u - u^k \rangle = 0.$$

Process (1.33) has quite a simple geometric interpretation. Set

$$H_k = \{u \in R^n \mid \langle \nabla f(u^k), u - u^k \rangle = -f(u^k)\}.$$

Note that the hyperplane H_k separates u^k and the solution set of problem
(1.32). In fact, we have

$$\langle \nabla f(u^k), u^k - u^k \rangle = 0 \geq -f(u^k),$$

the inequality being strict if $f(u^k) > 0$. On the other hand, if $f(u^*) = 0$, then

$$\langle \nabla f(u^k), u^* - u^k \rangle \leq f(u^*) - f(u^k) = -f(u^k)$$

due to Lemma 1.1.3. It is easy to see that u^{k+1} in (1.33) is the projection of u^k onto the hyperplane H_k, so that the distance from u^{k+1} to each solution of (1.32) has to decrease in comparison with that from u^k. More precisely, we have

$$\|u^{k+1} - u^*\|^2 \leq \|u^k - u^*\|^2 - (f(u^k)/\|\nabla f(u^k)\|)^2, \qquad (1.34)$$

where u^* is any solution of (1.32).

We now consider an extension of this approach to the case of VI (1.1).

Definition 1.2.2. Let W be a nonempty closed set in R^n. A mapping $P : R^n \to R^n$ is said to be
(a) *feasible with respect to W*, if

$$P(u) \in W, \qquad \forall u \in R^n;$$

(b) *quasi-nonexpansive with respect to W*, if for every $u \in R^n$, we have

$$\|P(u) - w\| \leq \|u - w\| \quad \forall w \in W.$$

We denote by $\mathcal{F}(W)$ the class of feasible quasi-nonexpansive mappings with respect to W.

Let us consider an iteration sequence $\{u^k\}$ generated in accordance with the following rules:

$$u^{k+1} := P_k(\tilde{u}^{k+1}), \tilde{u}^{k+1} := u^k - \gamma_k \sigma_k g^k, P_k \in \mathcal{F}(U), \qquad (1.35)$$
$$\gamma_k \in [0,2], \langle g^k, u^k - u^* \rangle \geq \sigma_k \|g^k\|^2 \geq 0 \qquad \forall u^* \in U^d. \quad (1.36)$$

It is easy to see that $\tilde{u}^{k+1}$ is the projection of u^k onto the hyperplane

$$H_k(\gamma_k) = \{v \in R^n \mid \langle g^k, v - u^k \rangle = -\gamma_k \sigma_k \|g^k\|^2\},$$

and that $H_k(1)$ separates u^k and U^d. Although $H_k(\gamma_k)$, generally speaking, does not possess this property, the distance from $\tilde{u}^{k+1}$ to each point of U^d cannot increase and the same is true for u^{k+1} since $U^d \subseteq U$. We now give the key property of the process (1.35), (1.36), which specifies the above statement.

Lemma 1.2.3. *Let a point u^{k+1} be chosen by (1.35), (1.36). Then we have*

$$\|u^{k+1} - u^*\|^2 \leq \|u^k - u^*\|^2 - \gamma_k(2 - \gamma_k)(\sigma_k \|g^k\|)^2 \qquad \forall u^* \in U^d. \quad (1.37)$$

Proof. Take any $u^* \in U^d$. By the properties of P_k and (1.36), we have

$$
\begin{aligned}
\|u^{k+1} - u^*\|^2 &\leq \|\tilde{u}^{k+1} - u^*\|^2 = \|u^k - \gamma_k \sigma_k g^k - u^*\|^2 \\
&= \|u^k - u^*\|^2 - 2\gamma_k \sigma_k \langle g^k, u^k - u^* \rangle + (\gamma_k \sigma_k \|g^k\|)^2 \\
&\leq \|u^k - u^*\|^2 - 2\gamma_k (\sigma_k \|g^k\|)^2 + (\gamma_k \sigma_k \|g^k\|)^2 \\
&= \|u^k - u^*\|^2 - \gamma_k (2 - \gamma_k)(\sigma_k \|g^k\|)^2,
\end{aligned}
$$

i.e. (1.37) is fulfilled, as desired.   □

It is obvious that (1.37) extends (1.34). The following assertions follow immediately from (1.37).

Lemma 1.2.4. *Let a sequence $\{u^k\}$ be constructed in accordance with the rules (1.35), (1.36). Then:*

(i) $\{u^k\}$ is bounded.

(ii) $\sum\limits_{k=0}^{\infty} \gamma_k (2 - \gamma_k)(\sigma_k \|g^k\|)^2 < \infty$.

(iii) For each limit point u^* of $\{u^k\}$ such that $u^* \in U^d$ we have

$$
\lim_{k \to \infty} u^k = u^*.
$$

Note that the sequence $\{u^k\}$ has limit points due to (i). Thus, due to (iii), it suffices to show that there exists the limit point of $\{u^k\}$ which belongs to U^d.

However, process (1.35), (1.36) is only a conceptual scheme, since it does not contain the rules of choosing the parameters g^k and σ_k satisfying (1.36). We now describe a basic scheme for constructing iterative methods, which can be made implementable in rather a simple manner.

Basic Scheme I. Choose a point $u^0 \in U$ and a sequence $\{\gamma_k\}$ such that

$$
\gamma_k \in [0, 2], k = 0, 1, \ldots; \quad \sum_{k=0}^{\infty} \gamma_k (2 - \gamma_k) = \infty. \tag{1.38}
$$

Also, choose a sequence of mappings $\{P_k\}$, where $P_k \in \mathcal{F}(U)$ for $k = 0, 1, \ldots$ At the kth iteration, $k = 0, 1, \ldots$, we have a point $u^k \in U$.

Step 1: Apply Procedure D_k and set $v^k := D_k(u^k)$. If $G(v^k) = 0$, stop.

Step 2: Set

$$
g^k := G(v^k), \sigma_k := \langle G(v^k), u^k - v^k \rangle / \|g^k\|^2, u^{k+1} := P_k(u^k - \gamma_k \sigma_k g^k).
$$

The iteration has been completed.

According to this description, the basic scheme contains the auxiliary procedure D_k which is used for finding a point $v^k \in U$ such that

$$
\langle G(v^k), u^k - v^k \rangle > 0
$$

when $u^k \notin U^*$. If $U^d \neq \emptyset$, this property together with (1.2) then implies that for every $u^* \in U^d$, we have

$$
\begin{aligned}
\langle g^k, u^k - u^* \rangle &= \langle G(v^k), u^k - v^k \rangle + \langle G(v^k), v^k - u^* \rangle \\
&\geq \langle G(v^k), u^k - v^k \rangle = \sigma_k \|g^k\|^2 \geq 0,
\end{aligned}
$$

i.e., (1.36) holds. Therefore, all the assertions of Lemmas 1.2.3 and 1.2.4 will be true for a sequence $\{u^k\}$ generated by the basic scheme, so that Lemma 1.2.4 (ii) and (1.38) then imply

$$
\liminf_{k \to \infty}(\sigma_k \|g^k\|) = 0.
$$

If this condition is sufficient for a limit point of $\{u^k\}$ to be in U^d, from Lemma 1.2.4 (iii) it then follows that $\{u^k\}$ converges to a solution of VI (1.1) since $U^d \subseteq U^*$ due to Proposition 1.1.2 (ii).

In order to implement Procedure D_k satisfying the conditions above we propose to make use of a solution of problem (1.25). Thus, we obtain a two-level iterative method which is convergent if there exists a solution to DVI (1.2). This assumption is rather mild: due to Proposition 1.1.2, $U^* = U^d$ if G is pseudomonotone, whereas the direct extensions of differentiable optimization methods, described in the previous subsection, are not convergent even if G is monotone.

We now consider the problem of implementing feasible quasi-nonexpansive mappings. To this end, we give several projection properties.

Proposition 1.2.1. *Suppose W is a nonempty convex closed set in R^n, u is an arbitrary point in R^n. Then:*

(i) There exists the unique projection $v = \pi_W(u)$ of the point u onto the set W.

(ii) A point $v \in W$ is a projection of u onto W, if and only if

$$
\langle v - u, w - v \rangle \geq 0 \quad \forall w \in W. \tag{1.39}
$$

(iii) The projection mapping $\pi_W(\cdot)$ is non-expansive.

Proof. It is clear that the point $v = \pi_W(u)$ is a solution of the following convex optimization problem:

$$
\min \to \{\varphi(w) \mid w \in W\},
$$

where $\varphi(w) = 0.5\|w - u\|^2$. Due to Theorem 1.1.1, this problem is equivalent to the variational inequality (1.39) with the underlying mapping $F(w) = w - u$, which is continuous and strongly monotone due to Proposition 1.1.9. From Proposition 1.1.4 (ii) it now follows that both problems have a unique solution, i.e., assertions (i) and (ii) hold. Next, take arbitrary $u', u'' \in R^n$ and set $v' = \pi_W(u')$, $v'' = \pi_W(u'')$. Applying (1.39) with $u = u'$, $v = v'$, $w = v''$ and with $u = u''$, $v = v''$, $w = v'$, respectively, gives

$$\langle v' - u', v'' - v' \rangle \geq 0$$

and

$$\langle v'' - u'', v' - v'' \rangle \geq 0.$$

Adding these inequalities and applying the Cauchy-Schwarz inequality, we obtain

$$\langle v'' - v', v'' - v' \rangle \leq \langle u'' - u'', v'' - v' \rangle \leq \|u'' - u'\| \|v'' - v'\|,$$

i.e.,

$$\|v'' - v'\| \leq \|u'' - u'\|.$$

Therefore, the mapping $\pi_W(\cdot)$ is non-expansive.   □

Thus, in Basic Scheme I, we can take the projection mapping $\pi_U(\cdot)$ as P_k. However, if the definition of the set U includes functional constraints, then the projection onto U cannot be found by a finite procedure. Nevertheless, in that case there exist finite procedures of finding the corresponding point $P_k(u)$, which are described in detail in Section 4.1.

1.3 Implementable CR Methods

In this section, we describe two CR methods and establish their convergence and some rates of convergence. The blanket assumptions of this section are the following.

Hypothesis (H1.3)

(a) U is a nonempty, closed and convex subset of R^n;
(b) V is a closed convex subset of R^n such that $U \subseteq V$;
(c) $G : V \to R^n$ is a locally Lipschitz continuous mapping;
(d) $U^d \neq \emptyset$.

1.3.1 A CR Method and its Properties

We now describe a CR method which follows Basic Scheme I and involves a solution of the auxiliary problem (1.25). We will show that our method converges to a solution of VI (1.1) under Hypothesis (H1.3).

Method 1.1. *Step 0 (Initialization)*: Choose a point $u^0 \in U$, a sequence $\{\gamma_k\}$ such that

$$\gamma_k \in [0,2], k = 0, 1, \dots; \quad \sum_{k=0}^{\infty} \gamma_k(2 - \gamma_k) = \infty. \tag{1.40}$$

Also, choose a family of mappings $\{T_k : U \times U \to R^n\}$ satisfying Assumptions (A1) – (A3) of Section 1.2 with $V = U$ and a sequence of mappings $\{P_k\}$, where $P_k \in \mathcal{F}(U)$ for $k = 0, 1, \dots$ Choose numbers $\alpha \in (0,1)$, $\beta \in (0,1)$, $\tilde{\theta} > 0$. Set $k := 0$.
  Step 1 (Auxiliary procedure D_k):
      Step 1.1 : Solve the auxiliary variational inequality problem of finding $z^k \in U$ such that

$$\langle G(u^k) + T_k(u^k, z^k), v - z^k \rangle \geq 0 \quad \forall v \in U \tag{1.41}$$

and set $p^k := z^k - u^k$. If $p^k = 0$, stop.
      Step 1.2: Determine m as the smallest number in Z_+ such that

$$u^k + \beta^m \tilde{\theta} p^k \in U, \langle G(u^k + \beta^m \tilde{\theta} p^k), p^k \rangle \leq \alpha \langle G(u^k), p^k \rangle, \tag{1.42}$$

set $\theta_k := \beta^m \tilde{\theta}$, $v^k := u^k + \theta_k p^k$. If $G(v^k) = 0$, stop.
      Step 2 (Main iteration): Set

$$g^k := G(v^k), \sigma_k := \langle G(v^k), u^k - v^k \rangle / \|g^k\|^2, u^{k+1} := P_k(u^k - \gamma_k \sigma_k g^k), \tag{1.43}$$

$k := k + 1$ and go to Step 1.

According to the description, at each iteration we solve the auxiliary problem (1.25) with $\lambda_k = 1$ and carry out an Armijo-Goldstein type linesearch procedure. Thus, our method requires no a priori information about the original problem (1.1). In particular, it does not use the Lipschitz constant for G.

From Lemma 1.2.1 we obtain the following immediately.

Lemma 1.3.1. *(i) Problem (1.41) has a unique solution.*
(ii) It holds that

$$\langle G(u^k), u^k - z^k \rangle \geq \langle T_k(u^k, z^k), z^k - u^k \rangle \geq \tau_k' \|z^k - u^k\|^2. \qquad (1.44)$$

(iii) The solution z^k of (1.41) coincides with $u^k \in U$ if and only if $u^k \in U^*$.

It follows from (1.42), (1.43), and (1.44) that $\sigma_k > 0$ when $u^k \notin U^*$. For each $u^* \in U^d$, we then have

$$\begin{aligned}
\langle g^k, u^k - u^* \rangle &= \langle G(v^k), u^k - v^k \rangle + \langle G(v^k), v^k - u^* \rangle \qquad (1.45) \\
&\geq \sigma_k \|g^k\|^2 \geq 0,
\end{aligned}$$

i.e., (1.36) holds, hence, all the assertions of Lemmas 1.2.3 and 1.2.4 are true for Method 1.1. We give them here for the convenience of the reader.

Lemma 1.3.2. *It holds that:*
(i)

$$\|u^{k+1} - u^*\|^2 \leq \|u^k - u^*\|^2 - \gamma_k(2 - \gamma_k)(\sigma_k\|g^k\|)^2 \qquad \forall u^* \in U^d.$$

(ii) $\{u^k\}$ is bounded.
(iii) $\sum_{k=0}^{\infty} \gamma_k(2 - \gamma_k)(\sigma_k\|g^k\|)^2 < \infty$.
(iv) For each limit point u^* of $\{u^k\}$ such that $u^* \in U^d$ we have

$$\lim_{k \to \infty} u^k = u^*.$$

In addition to Assumptions (A1) – (A3) of Section 1.2 we consider the following one.
(A4) *It holds that*

$$0 < \tau' \leq \tau_k' \leq \tau_k'' \leq \tau'' < \infty. \qquad (1.46)$$

Assumption (A4) is not too restrictive. Obviously, it is fulfilled if the mappings $T_k(u, \cdot)$ are Lipschitz continuous with constant τ'' and strongly monotone with constant τ'. For the case of (1.27), it holds when $\|A_k\| \leq \tau''$ and

$$\tau'\|p\|^2 \leq \langle A_k p, p \rangle \quad \forall p \in R^n.$$

Lemma 1.3.3. *Suppose that (1.46) holds. Then:*
   (i) $\{z^k\}$ is bounded.
   (ii) $\theta_k \geq \theta' > 0$ for $k = 0, 1, \ldots$

Proof. From (1.44), (1.46) and the Cauchy - Schwarz inequality it follows that

$$\|G(u^k)\| \geq \tau' \|u^k - z^k\|.$$

Since $\{u^k\}$ is bounded due to Lemma 1.3.2 (ii) and G is locally Lipschitz continuous, the above inequality implies that $\{z^k\}$ is also bounded and that assertion (i) is true. Next, there exists a convex bounded subset W of U such that $\{u^k\}$ and $\{z^k\}$ are contained in W. Then G is Lipschitz continuous on W. It follows that for each $\theta \in (0, 1]$,

$$
\begin{aligned}
\langle G(u^k + \theta p^k), p^k \rangle &= \langle G(u^k), p^k \rangle + \langle G(u^k + \theta p^k) - G(u^k), p^k \rangle \\
&\leq \langle G(u^k), p^k \rangle + L\theta \|p^k\|^2,
\end{aligned}
$$

where L is the Lipschitz constant for G on W. Applying (1.44) and (1.46) in the above inequality gives

$$
\begin{aligned}
\langle G(u^k + \theta p^k), p^k \rangle &\leq (1 - \theta L/\tau_k')\langle G(u^k), p^k \rangle \leq (1 - \theta L/\tau')\langle G(u^k), p^k \rangle \\
&\leq \alpha \langle G(u^k), p^k \rangle
\end{aligned}
$$

when $\theta \leq \theta' = (1 - \alpha)\tau'/L$. Taking into account (1.42) we see that (ii) holds with $\theta' = \min\{\bar{\theta}, \beta\theta', 1\}$.   □

Thus, according to Lemma 1.3.3 (ii), the linesearch procedure in Step 2 is always finite. Moreover, the proof of Lemma 1.3.3 shows that we can choose $\theta_k = \theta \leq (1 - \alpha)\tau'/L$ for all k. Then we need no line searches. Nevertheless, since the Lipschitz constant is usually difficult to compute, we make use of the Armijo-Goldstein type linesearch.

1.3.2 Convergence and Rates of Convergence

We are now in a position to establish the following convergence theorem for Method 1.1.

Theorem 1.3.1. *Let a sequence $\{u^k\}$ be generated by Method 1.1. Suppose (1.46) holds. Then:*
   (i) If the method terminates at Step 1.1 (Step 1.2) of the kth iteration, $u^k \in U^*$ ($v^k \in U^*$).
   (ii) If $\{u^k\}$ is infinite, there exists a limit point u^* of $\{u^k\}$ which lies in U^*.
   (iii) If $\{u^k\}$ is infinite and

$$U^* = U^d, \tag{1.47}$$

we have

$$\lim_{k \to \infty} u^k = u^* \in U^*.$$

Proof. Assertion (i) is obviously true due to the stopping rule and Lemma 1.3.1 (iii). Next, from (1.42), (1.44) and Lemma 1.3.3 (ii) it follows that

$$
\begin{aligned}
\omega_k &= \langle G(v^k), u^k - v^k \rangle = \theta_k \langle G(v^k), u^k - z^k \rangle \\
&\geq \alpha \theta_k \langle G(u^k), u^k - z^k \rangle \geq \alpha \theta' \langle T_k(u^k, z^k), z^k - u^k \rangle, \quad (1.48)
\end{aligned}
$$

hence

$$
\omega_k \geq \alpha \theta' \langle T_k(u^k, z^k), z^k - u^k \rangle \geq \alpha \theta' \tau_k' \| z^k - u^k \|^2.
$$

Since $\{u^k\}$ and $\{z^k\}$ are bounded due to Lemma 1.3.2 (ii) and Lemma 1.3.3 (i), respectively, so are $\{v^k\}$ and $\{G(v^k)\}$. It then follows from (1.40), (1.46) and Lemma 1.3.2 (iii) that

$$
\liminf_{k \to \infty} \| z^k - u^k \| = 0.
$$

Hence, there exist subsequences $\{u^{k_s}\}$ and $\{z^{k_s}\}$ such that

$$
\lim_{s \to \infty} \| z^{k_s} - u^{k_s} \| = 0. \quad (1.49)
$$

Since $\{u^{k_s}\}$ and $\{z^{k_s}\}$ are bounded, they have limit points u^* and z^*, respectively. Without loss of generality, we suppose that

$$
\lim_{s \to \infty} u^{k_s} = u^*, \ \lim_{s \to \infty} z^{k_s} = z^*.
$$

These relations together with (1.49) yield

$$
u^* = z^* \in U. \quad (1.50)
$$

Next, by the definition of z^{k_s}, for every $v \in U$, we have

$$
\langle G(u^{k_s}), v - z^{k_s} \rangle + \langle T_{k_s}(u^{k_s}, z^{k_s}), v - z^{k_s} \rangle \geq 0,
$$

i.e.,

$$
\langle G(u^{k_s}), v - z^{k_s} \rangle \geq -\| T_{k_s}(u^{k_s}, z^{k_s}) \| \| v - z^{k_s} \|. \quad (1.51)
$$

From (1.46) and (1.49) it follows that

$$
\limsup_{s \to \infty} \| T_{k_s}(u^{k_s}, z^{k_s}) \| \leq \limsup_{s \to \infty} (\tau_{k_s}'' \| z^{k_s} - u^{k_s} \|) = 0.
$$

Hence, passing the limit $s \to \infty$ in (1.51) and using (1.50), we obtain

$$
\langle G(u^*), v - u^* \rangle \geq 0 \quad \forall v \in U,
$$

i.e., $u^* \in U^*$ and assertion (ii) holds. In case (1.47), we have $u^* \in U^d$ and assertion (iii) follows from Lemma 1.3.2 (iv). The proof is complete. □

Observe that the value $\langle T_k(u^k, z^k), z^k - u^k \rangle$ can be regarded as an error bound for Method 1.1 due to Lemma 1.3.1. Taking it as a basis we obtain a rate of convergence.

Theorem 1.3.2. *Let an infinite sequence* $\{u^k\}$ *be generated by Method 1.1. Suppose that (1.46) holds and that*

$$\gamma_k = \gamma \in (0,2), \quad k = 0, 1, \ldots \tag{1.52}$$

Then

$$\liminf_{k \to \infty} \left(\langle T_k(u^k, z^k), z^k - u^k \rangle \sqrt{k+1} \right) = 0. \tag{1.53}$$

Proof. By using the argument as in the proof of Theorem 1.3.1, we see that $\{G(v^k)\}$ is bounded, i.e.,

$$\|G(v^k)\| \le C < \infty \qquad k = 0, 1, \ldots, \tag{1.54}$$

besides, (1.48) holds, hence along with Lemma 1.3.3 (ii), we have

$$\langle G(v^k), u^k - v^k \rangle \ge \alpha \theta' \langle T_k(u^k, z^k), z^k - u^k \rangle. \tag{1.55}$$

From (1.52) and Lemma 1.3.2 (iii) it now follows that

$$\sum_{k=0}^{\infty} \langle T_k(u^k, z^k), z^k - u^k \rangle^2 < \infty.$$

If we now suppose that (1.53) does not hold, then there exists a number $\mu > 0$ such that $\langle T_k(u^k, z^k), z^k - u^k \rangle \ge \mu/\sqrt{k+1}$ and the above inequality then implies

$$\infty > \mu^2 \sum_{k=0}^{\infty} 1/(k+1),$$

which is a contradiction. The proof is complete. $\qquad \square$

Under additional assumptions the rate of convergence can be improved. Namely, let us consider the following assumption.

(A5) *Let* $\{T_k\}$ *be a sequence of mappings satisfying Assumptions (A1) – (A4) and let* $z_\lambda(u)$ *denote a (unique) solution to the problem*

$$\langle G(u) + \lambda^{-1} T_k(u, z_\lambda(u)), v - z_\lambda(u) \rangle \ge 0 \quad \forall v \in U, \tag{1.56}$$

with $0 < \lambda' \le \lambda \le \lambda'' < \infty$. *Then there exist a number* $\delta > 0$ *and a number* $\mu > 0$, *which can only depend on* δ, λ', *and* λ'', *such that for any point* $u \in V \supseteq U$, *the inequality*

$$\langle T_k(u, z_\lambda(u)), z_\lambda(u) - u \rangle \ge \mu \, d(u, U^*)^2 \tag{1.57}$$

holds whenever $\langle T_k(u, z_\lambda(u)), z_\lambda(u) - u \rangle \le \delta$.

Note that, if (1.47) holds, the set U^* must be convex and closed due to Proposition 1.1.2 (i), and there exists the projection $\pi_{U^*}(u)$ of every point $u \in R^n$ onto U^*, then $d(u, U^*) = \|u - \pi_{U^*}(u)\|$. We now describe some classes of problems which provide for Assumption (A5) to hold.

Proposition 1.3.1. *Let the mapping G be strongly monotone on U. Suppose also that at least one of the following assumptions holds:*
(i) the set V associated to Assumption (A5) is bounded;
(ii) G is Lipschitz continuous.
Then Assumption (A5) is fulfilled with $\delta = \infty$.

Proof. First we note that, under the assumption of strong monotonicity, VI (1.1) has a unique solution due to Proposition 1.1.4 (ii). We denote by u^* this solution. Then, from the definition of $z_\lambda(u)$ in (1.56) and the properties of G and T_k, we have

$$
\begin{aligned}
\tau \|z_\lambda(u) - u^*\|^2 &\leq \langle G(z_\lambda(u)) - G(u^*), z_\lambda(u) - u^* \rangle \\
&\leq \langle G(z_\lambda(u)), z_\lambda(u) - u^* \rangle \\
&= \langle G(z_\lambda(u)) - G(u) - \lambda^{-1} T_k(u, z_\lambda(u)), z_\lambda(u) - u^* \rangle \\
&+ \langle G(u) + \lambda^{-1} T_k(u, z_\lambda(u)), z_\lambda(u) - u^* \rangle \\
&\leq \langle G(z_\lambda(u)) - G(u), z_\lambda(u) - u^* \rangle \\
&- \lambda^{-1} \langle T_k(u, z_\lambda(u)) - T_k(u, u), z_\lambda(u) - u^* \rangle \\
&\leq (L + \tau''/\lambda) \|u - z_\lambda(u)\| \|z_\lambda(u) - u^*\|,
\end{aligned}
$$

where τ is the modulus of strong monotonicity, L is the Lipschitz constant for G on V. It follows that

$$
(L + \tau''/\lambda) \|u - z_\lambda(u)\|/\tau \geq \|z_\lambda(u) - u^*\| \geq \|u - u^*\| - \|u - z_\lambda(u)\|
$$

and that

$$
\langle T_k(u, z_\lambda(u)), z_\lambda(u) - u \rangle \geq \tau' \|z_\lambda(u) - u\|^2 \geq \mu \|u - u^*\|^2,
$$

where $\mu = \tau'/(1 + (L + \tau''/\lambda')/\tau)^2$. Therefore, (1.57) is fulfilled with $\delta = \infty$, as desired. $\square$

Proposition 1.3.2. *Suppose that U is a polyhedral set. Suppose also that at least one of the following assumptions holds:*
(i) G is affine;
(ii) $G(u) = E^T F(Eu) + q$, where E is an $m \times n$ matrix with no zero column, q is a vector in R^n and $F : R^m \to R^m$ is a strongly monotone Lipschitz continuous.
Then Assumption (A5) is fulfilled.

Since the proof of Proposition 1.3.2 need some additional properties we give it in Section 4.2.

We now establish rates of convergence for Method 1.1 under Assumption (A5). First we recall the well-known property of number sequences.

Lemma 1.3.4. *[42, Chapter 3, Lemma 1.4] Let η_k be a sequence of nonnegative numbers such that*

$$\eta_k - \eta_{k+1} \geq \kappa' \eta_k^2, \quad k = 0, 1, \ldots,$$

where $\kappa' > 0$. Then, there exists a number $\kappa'' < \infty$ such that

$$\eta_k \leq \kappa''/k, \quad k = 0, 1, \ldots$$

Theorem 1.3.3. *Let an infinite sequence $\{u^k\}$ be constructed by Method 1.1, where $\{\gamma_k\}$ is chosen in accordance with rule (1.52). Suppose that (1.47) and Assumption (A5) hold.*
Then:
(i) the sequence $\{\|u^k - \pi_{U^*}(u^k)\|\}$ converges to zero in the rate $O(1/\sqrt{k})$;
(ii) if $U = R^n$, $\{\|u^k - \pi_{U^*}(u^k)\|\}$ converges to zero in a linear rate.

Proof. Following the proof of Theorems 1.3.1 and 1.3.2, we see that (1.54) and (1.55) hold, too. Besides, from Lemma 1.3.2 (i), (1.47), (1.48) and (1.52) it follows that

$$\|u^{k+1} - \pi_{U^*}(u^{k+1})\|^2 \leq \|u^k - \pi_{U^*}(u^k)\|^2$$
$$- \gamma(2 - \gamma)(\alpha\theta'\langle T_k(u^k, z^k), z^k - u^k\rangle/\|G(v^k)\|)^2. \qquad (1.58)$$

Therefore, by (1.54), $\langle T_k(u^k, v^k), v^k - u^k\rangle \to 0$ as $k \to \infty$ and

$$\langle T_k(u^k, z^k), z^k - u^k\rangle \geq \mu \|u^k - \pi_{U^*}(u^k)\|^2 \qquad (1.59)$$

for k large enough due to Assumption (A5). Applying this inequality in (1.58), we see that the sequence $\{\|u^k - \pi_{U^*}(u^k)\|^2\}$ satisfies all the conditions of Lemma 1.3.4, i.e., assertion (i) is true.

Next, in the case of (ii), we must have

$$G(u^k) + T_k(u^k, z^k) = 0. \qquad (1.60)$$

Combining this property with (A1) – (A4) and with the fact that G is locally Lipschitz continuous gives

$$\|G(v^k)\| \leq \|G(u^k)\| + L\|v^k - u^k\| \leq \|T_k(u^k, z^k)\| + L\tilde{\theta}\|u^k - z^k\|$$
$$\leq (\tau'' + L\tilde{\theta})\|u^k - z^k\|,$$

where L is the corresponding Lipschitz constant. By (1.44), we then have

$$\|G(v^k)\|^2 \leq [(\tau'' + L\tilde{\theta})^2/\tau']\langle T_k(u^k, z^k), z^k - u^k\rangle.$$

Using this relation in (1.58) gives

$$\|u^{k+1} - \pi_{U^*}(u^{k+1})\|^2 \leq \|u^k - \pi_{U^*}(u^k)\|^2 - \nu\langle T_k(u^k, v^k), v^k - u^k\rangle,$$

where

$$\nu = \gamma(2 - \gamma)\tau'[\alpha\theta'/(\tau'' + L\tilde{\theta})]^2.$$

Similarly, $\langle T_k(u^k, v^k), v^k - u^k \rangle \to 0$ as $k \to \infty$ and (1.59) holds for k large enough. Therefore,

$$\lim_{k \to \infty} (\|u^{k+1} - \pi_{U^*}(u^{k+1})\|/\|u^k - \pi_{U^*}(u^k)\|) \le (1 - \nu)^{1/2} \in (0, 1),$$

and assertion (ii) follows.   □

We now give conditions that ensure finite termination of the method. Namely, let us consider the following assumption.

(A6) *There exists a number $\mu' > 0$ such for each point $u \in U$, the following inequality holds:*

$$\langle G(u), u - \pi_{U^*}(u) \rangle \ge \mu' \, \|u - \pi_{U^*}(u)\|. \tag{1.61}$$

Theorem 1.3.4. *Let a sequence $\{u^k\}$ be constructed by Method 1.1. Suppose that (1.47) and Assumptions (A4) and (A6) hold. Then the method terminates with a solution.*

Proof. Assume for contradiction that $\{u^k\}$ is infinite. Then, following the proof of Theorems 1.3.1 and 1.3.2, we see that (1.54) and (1.55) hold. Then, by (1.44), (1.40), and Lemma 1.3.2 (iii), we have

$$\liminf_{k \to \infty} \|z^k - u^k\| = 0. \tag{1.62}$$

Next, on account of (1.41), Assumptions (A1) – (A4), (A6), and Lemma 1.3.3 (i), we obtain

$$
\begin{aligned}
\mu'\|z^k - \pi_{U^*}(z^k)\| \ &\le \ \langle G(z^k), z^k - \pi_{U^*}(z^k) \rangle \\
&= \ \langle G(z^k) - G(u^k) - T_k(u^k, z^k), z^k - \pi_{U^*}(z^k) \rangle \\
&+ \ \langle G(u^k) + T_k(u^k, z^k), z^k - \pi_{U^*}(z^k) \rangle \\
&\le \ (L\|z^k - u^k\| + \tau''\|z^k - u^k\|)\|z^k - \pi_{U^*}(z^k)\|,
\end{aligned}
$$

where L is the corresponding Lipschitz constant for G. It follows that

$$\|u^k - z^k\| \ge \mu'/(L + \tau''),$$

which contradicts (1.62). The proof is complete.   □

It should be noted that Assumption (A6) can be viewed as an extension of the well-known sharp optimum condition in optimization; e.g., see [179, Chapters 7 and 10] and [54]. Indeed, this latter condition can be written as:

$$\text{for every } u \in U, \quad f(u) - f(u^*(u)) \ge \mu\|u - u^*(u)\|, \tag{1.63}$$

where $u^*(u)$ is the projection of u onto the solution set U_f of the problem

$$\min \to \{f(u) \mid u \in U\}.$$

Suppose that $f : R^n \to R$ is convex and differentiable, then U_f is convex and closed due to Theorem 1.1.1 and Propositions 1.1.2 and 1.1.9, hence for each $u \in R^n$ there exists the unique projection $u^*(u)$. If one now sets $G(u) = \nabla f(u)$, then, due to Theorem 1.1.1, VI (1.1) becomes equivalent to the optimization problem above, moreover, (1.63) implies (A6). In fact, by Lemma 1.1.3, for every $u \in U$, one has

$$\langle \nabla f(u), u - u^*(u) \rangle \geq f(u) - f(u^*(u)) \geq \mu \|u - u^*(u)\|.$$

1.3.3 A Modified Line Search

In Step 1 of Method 1.1, we solved the auxiliary problem (1.41), which corresponds to (1.25) with $\lambda_k = 1$, and afterwards found the stepsize along the ray $u^k + \theta(z^k - u^k)$. However, it is clear that one can satisfy condition (1.42) by sequentially solving problem (1.25) with various λ_k. We now describe a CR method which involves a modified linesearch procedure.

Method 1.2. *Step 0 (Initialization)*: Choose a point $u^0 \in U$, a sequence $\{\gamma_k\}$ satisfying (1.40), a family of mappings $\{T_k : U \times U \to R^n\}$ satisfying Assumptions (A1) – (A4) with $V = U$ and a sequence of mappings $\{P_k\}$, where $P_k \in \mathcal{F}(U)$ for $k = 0, 1, \dots$ Choose numbers $\alpha \in (0,1)$, $\beta \in (0,1)$, $\tilde{\theta} > 0$. Set $k := 0$.

Step 1 (Auxiliary procedure D_k):

   Step 1.1: Find m as the smallest number in Z_+ such that

$$\langle G(z^{k,m}), z^{k,m} - u^k \rangle \leq \alpha \langle G(u^k), z^{k,m} - u^k \rangle, \tag{1.64}$$

where $z^{k,m}$ is a solution of the auxiliary problem of finding $\bar{z} \in U$ such that

$$\langle G(u^k) + (\tilde{\theta}\beta^m)^{-1} T_k(u^k, \bar{z}), u - \bar{z} \rangle \geq 0 \quad \forall u \in U. \tag{1.65}$$

   Step 1.2: Set $\theta_k := \beta^m \tilde{\theta}$, $v^k := z^{k,m}$. If $u^k = v^k$ or $G(v^k) = 0$, stop.

Step 2 (Main iteration): Set

$$g^k := G(v^k), \sigma_k := \langle G(v^k), u^k - v^k \rangle / \|g^k\|^2, \tag{1.66}$$

$$u^{k+1} := P_k(u^k - \gamma_k \sigma_k g^k), \tag{1.67}$$

$k := k + 1$ and go to Step 1.

To establish convergence of Method 1.2 we need analogues of Lemmas 1.3.1 – 1.3.3.

Lemma 1.3.5. *(i) Problem (1.65) has a unique solution.*
(ii) It holds that

$$(\tilde{\theta}\beta^m)\langle G(u^k), u^k - z^{k,m} \rangle \geq \langle T_k(u^k, z^{k,m}), z^{k,m} - u^k \rangle$$
$$\geq \tau_k' \|z^{k,m} - u^k\|^2 \quad \forall m \in Z_+. \tag{1.68}$$

(iii) The solution $z^{k,m}$ of (1.65) coincides with $u^k \in U$ if and only if $u^k \in U^*$.
(iv) The assertions of Lemma 1.3.2 remain valid for Method 1.2.

Proof. Parts (i) – (iii) follow directly from Lemma 1.2.1. Next, from (1.68) and (1.66) we see that $\sigma_k > 0$ when $u^k \notin U^*$ and that (1.36) holds for each $u^* \in U^d$ due to (1.45). Therefore, part (iv) now follows from Lemmas 1.2.3 and 1.2.4. $\qquad\qquad\qquad\qquad\qquad\qquad\qquad\qquad\qquad\qquad\qquad\qquad\qquad\qquad$ □

Lemma 1.3.6. *(i) $\{v^k\}$ is bounded.*
  (ii) $\theta_k \geq \theta' > 0$ for $k = 0, 1, \ldots$

Proof. From (1.46), (1.68) and the Cauchy - Schwarz inequality it follows that

$$\tilde{\theta}\|G(u^k)\| \geq \tau'\|u^k - z^{k,m}\| \quad \forall m \in Z_+.$$

Since $\{u^k\}$ is bounded due to Lemmas 1.3.5 (iv) and 1.3.2 (ii), the above inequality now implies that so is $\{v^k\}$ and that assertion (i) is true. Next, there exists a convex bounded subset W of U such that $\{u^k\}$ and $\{z^{k,m}\}$ are contained in W. Then G is Lipschitz continuous on W. It follows that

$$
\begin{aligned}
\langle G(z^{k,m}), z^{k,m} - u^k \rangle &\leq \langle G(u^k), z^{k,m} - u^k \rangle \\
&+ \langle G(z^{k,m}) - G(u^k), z^{k,m} - u^k \rangle \\
&\leq \langle G(u^k), z^{k,m} - u^k \rangle + L\|z^{k,m} - u^k\|^2,
\end{aligned}
$$

where L is the Lipschitz constant for G on W. Applying (1.46) and (1.68) in the above inequality gives

$$
\begin{aligned}
\langle G(z^{k,m}), z^{k,m} - u^k \rangle &\leq (1 - (\tilde{\theta}\beta^m)L/\tau')\langle G(u^k), z^{k,m} - u^k \rangle \\
&\leq \alpha \langle G(u^k), z^{k,m} - u^k \rangle
\end{aligned}
$$

when $(1 - (\tilde{\theta}\beta^m)L/\tau') \geq \alpha$ or equivalently, $(\tilde{\theta}\beta^m) \leq (1 - \alpha)\tau'/L$. Taking into account (1.64), we see that part (ii) holds with $\theta' = \min\{\tilde{\theta}, (1 - \alpha)\tau'/L\}$. □

We now establish a convergence result for Method 1.2.

Theorem 1.3.5. *Let a sequence $\{u^k\}$ be generated by Method 1.2. Then:*
  (i) If the method terminates at the kth iteration, $v^k \in U^*$.
  (ii) If $\{u^k\}$ is infinite, there exists a limit point u^* of $\{u^k\}$ which lies in U^*.
  (iii) If $\{u^k\}$ is infinite, and (1.47) holds, we have

$$\lim_{k \to \infty} u^k = u^* \in U^*.$$

Proof. Assertion (i) is obviously true due to the stopping rule and Lemmas 1.3.5 (iii) and 1.3.6 (ii). Next, let $\{u^k\}$ be infinite. Then, taking into account (1.64), (1.68) and Lemma 1.3.6 (ii), we have

$$
\begin{aligned}
\omega_k &= \langle G(v^k), u^k - v^k \rangle \geq \alpha \langle G(u^k), u^k - v^k \rangle \geq \alpha \theta_k^{-1} \langle T_k(u^k, v^k), v^k - u^k \rangle \\
&\geq \alpha(\tau_k'/\theta_k)\|v^k - u^k\|^2 \geq \alpha(\tau'/\theta')\|v^k - u^k\|^2. \tag{1.69}
\end{aligned}
$$

Since $\{u^k\}$ and $\{v^k\}$ are bounded due to Lemmas 1.3.5 (iv), 1.3.2 (ii) and 1.3.6 (i), so is $\{G(v^k)\}$. It then follows from (1.40) and Lemmas 1.3.5 (iv), 1.3.2 (iii) that

$$\liminf_{k\to\infty} \|v^k - u^k\| = 0. \tag{1.70}$$

The sequences $\{u^k\}$ and $\{v^k\}$ must have limit points. Without loss of generality we can suppose that there exist subsequences $\{u^{k_s}\}$ and $\{v^{k_s}\}$ such that

$$\lim_{s\to\infty} u^{k_s} = \lim_{s\to\infty} v^{k_s} = u^* \in U. \tag{1.71}$$

From (1.65) with $\bar{z} = v^{k_s}$, Lemma 1.3.6 (ii), and the properties of T_k, for every $u \in U$, we have

$$\begin{aligned}
\langle G(u^{k_s}), u - v^{k_s} \rangle &\geq -\|T_{k_s}(u^{k_s}, v^{k_s})\|\|u - v^{k_s}\|/\theta_{k_s} \\
&\geq -\tau''\|u^{k_s} - v^{k_s}\|\|u - v^{k_s}\|/\theta'.
\end{aligned}$$

Taking the limit $s \to \infty$ in this inequality and taking into account (1.71), we obtain

$$\langle G(u^*), u - u^* \rangle \geq 0 \quad \forall u \in U,$$

i.e., $u^* \in U^*$. This proves part (ii). Part (iii) follows from Lemmas 1.3.5 (iv) and 1.3.2 (iv). □

The convergence rates of Method 1.1, which were obtained in Theorems 1.3.2 – 1.3.4, are in general true for Method 1.2.

Theorem 1.3.6. *Let an infinite sequence $\{u^k\}$ be generated by Method 1.2, where $\{\gamma_k\}$ is chosen in accordance with rule (1.52). Then*

$$\liminf_{k\to\infty} \left(\langle T_k(u^k, v^k), v^k - u^k \rangle \sqrt{k+1} \right) = 0.$$

The proof is similar to that of Theorem 1.3.2 with using (1.69) instead of (1.48).

Theorem 1.3.7. *Let an infinite sequence $\{u^k\}$ be constructed by Method 1.2, where $\{\gamma_k\}$ is chosen in accordance with rule (1.52). Suppose that (1.47) and Assumption (A5) hold. Then:*
(i) the sequence $\{\|u^k - \pi_{U^*}(u^k)\|\}$ converges to zero in the rate $O(1/\sqrt{k})$;
(ii) if $U = R^n$, $\{\|u^k - \pi_{U^*}(u^k)\|\}$ converges to zero in a linear rate.

Proof. The assertions above are proved by the argument similar to that in the proof of Theorem 1.3.3 with using (1.68) and (1.69) instead of (1.44) and (1.48), respectively. Besides, in case (ii), we make use of

$$G(u^k) + \theta_k^{-1} T_k(u^k, v^k) = 0$$

instead of (1.60). □

Theorem 1.3.8. *Let a sequence $\{u^k\}$ be constructed by Method 1.2. Suppose that (1.47) and Assumption (A6) hold. Then the method terminates with a solution.*

Proof. Assume for contradiction that $\{u^k\}$ is infinite. Then, following the proof of Theorem 1.3.5, we see that (1.70) holds. On the other hand, from (1.65), Assumptions (A1) – (A4), (A6), Lemmas 1.3.5 (iv) and 1.3.6 it follows that

$$
\begin{aligned}
\mu' \|v^k - \pi_{U^*}(v^k)\| &\leq \langle G(v^k), v^k - \pi_{U^*}(v^k)\rangle \\
&= \langle G(v^k) - G(u^k) - \theta_k^{-1} T_k(u^k, v^k), v^k - \pi_{U^*}(v^k)\rangle \\
&+ \langle G(u^k) + \theta_k^{-1} T_k(u^k, v^k), v^k - \pi_{U^*}(v^k)\rangle \\
&\leq (L\|v^k - u^k\| + \tau_k'' \theta_k^{-1} \|v^k - u^k\|)\|v^k - \pi_{U^*}(v^k)\| \\
&\leq (L + \tau''/\theta')\|v^k - u^k\|)\|v^k - \pi_{U^*}(v^k)\|.
\end{aligned}
$$

Therefore, we have

$$ \|v^k - u^k\| \geq \mu'/(L + \tau''/\theta'), $$

which contradicts (1.70).   □

Thus, Method 1.2 has the same rates of convergence as those of Method 1.1. However, the auxiliary procedure D_k in Method 1.2 involves, generally speaking, the auxiliary VI (1.65) to be solved several times, whereas the auxiliary problem (1.41) in Step 1 of Method 1.1 is solved once. In the simplest case, where rule (1.27) is applied with $A_k \equiv I$, these auxiliary procedures can be viewed as extensions of different variants of the gradient projection iterations (e.g., see [42, Chapter 3]).

1.3.4 Acceleration Techniques for CR Methods

In Lemma 1.3.2 (i), we obtained the following key property of Methods 1.1 and 1.2:

$$ \|u^{k+1} - u^*\|^2 \leq \|u^k - u^*\|^2 - \gamma_k(2 - \gamma_k)(\sigma_k\|g^k\|)^2 \quad \forall u^* \in U^d, $$

where

$$ \sigma_k\|g^k\| = \langle G(v^k), u^k - v^k\rangle/\|G(v^k)\|. $$

Hence, one can accelerate the convergence of the sequence $\{u^k\}$ to a solution by increasing the term $\sigma_k\|g^k\|$, or equivalently, by tending the value $\|g^k\|$ to zero as $u^k \to u^*$. With this purpose, one may make use of the additional projection operation of g^k onto a linear subspace which is tangent to the set U at u^k. More precisely, fix an iteration number k. Let Q_k be an $m_k \times n$ matrix with linearly independent rows $q^{k,i}, i = 1, \ldots, m_k; m_k \leq n$. Set

$$ S_k = \{u \in R^n \mid Q_k u = 0\}, $$

then the projection of an element $g \in R^n$ onto the linear subspace S_k is defined as follows:

$$\pi_{S_k}(g) = (I - Q_k^T(Q_kQ_k^T)^{-1}Q_k)g; \qquad (1.72)$$

e.g., see [175, Section 4.5].

The first modification of the methods, denoted by M1, consists of replacing rule (1.66) with the following

$$g_k := \pi_{S_k}(G(v^k)), \sigma_k := \langle G(v^k), u^k - v^k \rangle / \|g^k\|^2. \qquad (1.73)$$

Nevertheless, we now need to specify the choice of Q_k. We first give the conditions which maintain convergence properties of the methods.

Proposition 1.3.3. *If*

$$\langle (Q_kQ_k^T)^{-1}Q_kG(v^k), Q_k(u^k - u^*) \rangle \leq 0 \quad \forall u^* \in U^d, \qquad (1.74)$$

then all the assertions of Theorems 1.3.1 – 1.3.4 (respectively, Theorems 1.3.5 – 1.3.8) remain valid for Method 1.1 (respectively, Method 1.2) with modification M1.

Proof. First we note that, in the modified methods,

$$\|g^k\| \leq \|G(v^k)\|$$

due to the non-expansive projection property; see Proposition 1.2.1. Next, by using (1.73) and (1.74) instead of (1.43), we have

$$
\begin{aligned}
\langle g^k, u^k - u^* \rangle &= \langle G(v^k), u^k - u^* \rangle - \langle Q_k^T(Q_kQ_k^T)^{-1}Q_kG(v^k), u^k - u^* \rangle \\
&\geq \langle G(v^k), u^k - v^k \rangle - \langle (Q_kQ_k^T)^{-1}Q_kG(v^k), Q_k(u^k - u^*) \rangle \\
&\geq \langle G(v^k), u^k - v^k \rangle = \sigma_k\|g^k\|^2 \geq 0,
\end{aligned}
$$

for each $u^* \in U^d$, i.e., (1.36) also holds, hence the assertions of Lemmas 1.3.2 and 1.3.5 (iv) remain valid. Following the proofs of Theorems 1.3.1 – 1.3.8 we then conclude that all the assertions are true for the modified methods as well. □

Let us consider rule (1.74) in detail. Choose $\tilde{q}^{k,i}$ as linearly independent normal vectors to U at u^k, i.e.,

$$\langle \tilde{q}^{k,i}, u - u^k \rangle \leq 0 \quad \forall u \in U, \qquad (1.75)$$

It follows that

$$\langle \tilde{q}^{k,i}, u^* - u^k \rangle \leq 0 \quad \forall u^* \in U^d.$$

Besides, we select normal vectors $\tilde{q}^{k,i}$ which satisfy the following additional property:

$$\langle \tilde{q}^{k,i}, G(v^k) \rangle \leq 0 \qquad (1.76)$$

and denote by I_k the index set of these vectors, i.e., set $Q_k^T = (\tilde{q}^{k,i})_{i \in I_k}$. Note that the matrix $(Q_k Q_k^T)^{-1}$ is then positive definite, since the rows of Q_k are linearly independent. Nevertheless, to provide (1.74) we need additional assumptions. On the other hand, note that vectors $Q_k(u^* - u^k)$ and $Q_k G(v^k)$ have only non-positive components due to (1.75) and (1.76). Hence, if we choose Q_k so that $Q_k Q_k^T$ is diagonal, then $(Q_k Q_k^T)^{-1}$ will be also diagonal with non-negative entries and (1.74) will be satisfied. For instance, if the rows $q^{k,i}$ are orthogonal to each other, then $Q_k Q_k^T$ is obviously diagonal. We now consider several classes of problems which admit such a choice.

Let

$$U = \{u \in R^n \mid h_i(u) \le 0, i = 1, \ldots, t\}$$

where $h_i : R^n \to R$, $i = 1, \ldots, t$ are convex differentiable functions. For each $u \in R^n$, define the index set

$$I_\varepsilon(u) = \{i \mid h_i(u) \ge -\varepsilon\}, \ \varepsilon \ge 0.$$

Therefore, we can choose I_k as the subset of $I_\varepsilon(u^k)$ where $\tilde{q}^{k,i} = h_i'(u^k)$ satisfy (1.75) and (1.76). Since h_i is convex, we have

$$\varepsilon \ge h_i(u^*) - h_i(u^k) \ge \langle q^{k,i}, u^* - u^k \rangle$$

and

$$h_i(u^k + \lambda G(v^k)) + \varepsilon \ge h_i(u^k + \lambda G(v^k)) - h_i(u^k) \ge \lambda \langle \tilde{q}^{k,i}, G(v^k) \rangle,$$

hence, I_k is nonempty if the point $u^k + \lambda G(v^k)$ is feasible for some $\lambda > 0$. Next, the vectors $q^{k,i}$ are obviously orthogonal to each other, if U is the non-negative orthant (i.e., in the case of CP) or a parallelepiped in R^n. Otherwise, we can take I_k containing a unique element, then S_k is a hyperplane and $Q_k Q_k^T$ is a number, i.e., (1.74) is also satisfied.

In addition to modification M1, we can consider another modification, denoted by M2, which consists of replacing all the vectors of the form $G(u)$ in the methods with their orthogonal projections onto subspaces S_k in accordance with (1.72). This technique is still heuristic, but it might be useful in numerical implementation of the methods.

1.4 Modified Rules for Computing Iteration Parameters

In the previous section, two implementable algorithms following the general CR approach of Section 1.2 were described. In this section, we consider modified rules for computing the stepsize and descent direction in the main process. We will show that new methods are convergent under the same assumptions as those in Section 1.3, but their rates of convergence can be strengthened.

1.4.1 A Modified Rule for Computing the Stepsize

We now describe another CR method with a modified rule of choosing the stepsize σ_k. It can be regarded as a modification of Method 1.2 and essentially exploits the projection properties. We suppose that Hypothesis $(H1.3)$ of Section 1.3 is fulfilled.

Method 1.3. *Step 0 (Initialization)*: Choose a point $u^0 \in U$, a sequence $\{\gamma_k\}$ satisfying (1.40), a family of mappings $\{T_k : U \times U \to R^n\}$ satisfying Assumptions (A1) – (A4) with $V = U$. Choose numbers $\alpha \in (0, 1)$, $\beta \in (0, 1)$, $\tilde{\theta} > 0$. Set $k := 0$.
   Step 1 (Auxiliary procedure D_k):
   Step 1.1 : Find m as the smallest number in Z_+ such that

$$\langle G(u^k) - G(z^{k,m}), u^k - z^{k,m} \rangle \leq (1 - \alpha)(\tilde{\theta}\beta^m)^{-1}$$
$$\langle T_k(u^k, z^{k,m}), z^{k,m} - u^k \rangle, \quad (1.77)$$

where $z^{k,m}$ is a solution of the auxiliary problem of finding $\bar{z} \in U$ such that (1.65) holds.
   Step 1.2: Set $\theta_k := \beta^m \tilde{\theta}$, $v^k := z^{k,m}$. If $u^k = v^k$ or $G(v^k) = 0$, stop.
   Step 2 (Main iteration): Set

$$t^k := G(v^k) - G(u^k) - \theta_k^{-1} T_k(u^k, v^k), \quad (1.78)$$
$$g^k := G(v^k), \sigma_k := \alpha\theta_k^{-1}\langle T_k(u^k, v^k), v^k - u^k \rangle / \|t^k\|^2, \quad (1.79)$$
$$u^{k+1} := \pi_U(u^k - \gamma_k\sigma_k g^k), \quad (1.80)$$

$k := k + 1$ and go to Step 1.

Thus, Method 1.3 involves rules (1.78), (1.79) instead of (1.43) (or (1.66)) for choosing the stepsize parameter σ_k, however, the descent direction is the same as that in Methods 1.1 and 1.2. In addition, rule (1.80) differs from (1.43) and (1.67), since it involves the projection operation onto U instead of making use of any feasible quasi-nonexpansive mapping P_k. It should be noted that assertions (i) - (iii) of Lemma 1.3.5 are true for Method 1.3 since it solves the same auxiliary problem (1.65). Moreover, from (1.77) – (1.79) it follows that

$$\begin{aligned}
\langle t^k, u^k - v^k \rangle &= \langle G(v^k) - G(u^k), u^k - v^k \rangle + \theta_k^{-1} \langle T_k(u^k, v^k), v^k - u^k \rangle \\
&\geq \left[-(1-\alpha)\theta_k^{-1} + \theta_k^{-1} \right] \langle T_k(u^k, v^k), v^k - u^k \rangle \\
&\geq \alpha \theta_k^{-1} \tau_k' \| u^k - v^k \|^2.
\end{aligned}$$

Hence, $u^k = v^k$ if $t^k = 0$ and rule (1.79) is well defined.

In order to establish convergence of Method 1.3 we need the following analogue of Lemma 1.3.2.

Lemma 1.4.1. *It holds that:*

(i)

$$\| u^{k+1} - u^* \|^2 \leq \| u^k - u^* \|^2 - \gamma_k(2 - \gamma_k)(\sigma_k \| t^k \|)^2 \qquad \forall u^* \in U^d.$$

(ii) $\{u^k\}$ *is bounded.*

(iii) $\sum_{k=0}^{\infty} \gamma_k(2 - \gamma_k)(\sigma_k \| t^k \|)^2 < \infty.$

(iv) For each limit point u^* of $\{u^k\}$ such that $u^* \in U^d$ we have

$$\lim_{k \to \infty} u^k = u^*.$$

Proof. First we show that for all k,

$$\sigma_k \geq 0, \langle g^k, u^k - u^* \rangle \geq \sigma_k \| t^k \|^2 + \langle g^k - t^k, u^k - u^{k+1} \rangle \quad \forall u^* \in U^d. \quad (1.81)$$

In fact, take any $u^* \in U^d$. Then, by (1.65), (1.77) – (1.80), we have

$$\begin{aligned}
&\langle g^k, u^k - u^* \rangle - \sigma_k \| t^k \|^2 - \langle g^k - t^k, u^k - u^{k+1} \rangle \\
\geq\ &\langle g^k, u^k - v^k \rangle - \sigma_k \| t^k \|^2 - \langle g^k - t^k, u^k - v^k \rangle \\
-\ &\langle G(u^k) + \theta_k^{-1} T_k(u^k, v^k), v^k - u^{k+1} \rangle \\
\geq\ &\langle G(v^k) - G(u^k), u^k - v^k \rangle - \theta_k^{-1} \langle T_k(u^k, v^k), u^k - v^k \rangle \\
+\ &\alpha \theta_k^{-1} \langle T_k(u^k, v^k), u^k - v^k \rangle \\
\geq\ &(1 - \alpha)\theta_k^{-1} \langle T_k(u^k, v^k), u^k - v^k \rangle \\
-\ &(1 - \alpha)\theta_k^{-1} \langle T_k(u^k, v^k), u^k - v^k \rangle = 0,
\end{aligned}$$

i.e., the second relation in (1.81) holds. The first relation in (1.81) follows from the properties of T_k.

For brevity, set $\tilde{u}^{k+1} := u^k - \gamma_k \sigma_k g^k$. By using (1.80) and the projection property (1.39) with $W = U$, $u = \tilde{u}^{k+1}$, $w = u^*$, we obtain

$$\begin{aligned}
\| u^{k+1} - u^* \|^2 &= \| u^{k+1} - \tilde{u}^{k+1} \|^2 \\
&+ 2\langle u^{k+1} - \tilde{u}^{k+1}, \tilde{u}^{k+1} - u^* \rangle + \| \tilde{u}^{k+1} - u^* \|^2 \\
&\leq \| \tilde{u}^{k+1} - u^* \|^2 - \| \tilde{u}^{k+1} - u^{k+1} \|^2 \\
&= \| u^k - u^* \|^2 - \| u^k - u^{k+1} \|^2 \\
&- 2\gamma_k \sigma_k (\langle g^k, u^k - u^* \rangle - \langle g^k, u^k - u^{k+1} \rangle).
\end{aligned}$$

Applying (1.40) and (1.81) gives

$$
\begin{aligned}
\|u^{k+1} - u^*\|^2 \ &\leq \ \|u^k - u^*\|^2 - 2\gamma_k\sigma_k(\sigma_k\|t^k\|^2 - \langle t^k, u^k - u^{k+1}\rangle) \\
&- \ \|u^k - u^{k+1}\|^2 = \|u^k - u^*\|^2 - \gamma_k(2-\gamma_k)(\sigma_k\|t^k\|)^2 \\
&- \ ((\gamma_k\sigma_k\|t^k\|)^2 - 2\gamma_k\sigma_k\langle t^k, u^k - u^{k+1}\rangle + \|u^k - u^{k+1}\|^2) \\
&= \ \|u^k - u^*\|^2 - \gamma_k(2-\gamma_k)(\sigma_k\|t^k\|)^2 \\
&- \ \|(u^k - u^{k+1}) - \gamma_k\sigma_k t^k\|^2 \\
&\leq \ \|u^k - u^*\|^2 - \gamma_k(2-\gamma_k)(\sigma_k\|t^k\|)^2.
\end{aligned}
$$

This proves part (i). Parts (ii) – (iv) follow directly from (i). □

We now consider the linesearch procedure in Step 1.

Lemma 1.4.2. *(i)* $\{v^k\}$ *is bounded.*
(ii) $\theta_k \geq \theta' > 0$ *for* $k = 0, 1, \ldots$

Proof. From (1.46), (1.68) and the Cauchy - Schwarz inequality it follows that

$$
\tilde{\theta}\|G(u^k)\| \geq \tau'\|u^k - z^{k,m}\| \quad \forall m \in Z_+.
$$

Since $\{u^k\}$ is bounded due to Lemma 1.4.1 (ii), the above inequality now implies that so is $\{v^k\}$ and that assertion (i) is true. Next, there exists a convex bounded subset W of U such that $\{u^k\}$ and $\{z^{k,m}\}$ are contained in W. Then G is Lipschitz continuous on W. It follows that

$$
\langle G(z^{k,m}), z^{k,m} - u^k\rangle \leq \langle G(u^k), z^{k,m} - u^k\rangle + L\|z^{k,m} - u^k\|^2,
$$

where L is the Lipschitz constant for G on W. Applying (A1) – (A4) in the above inequality gives

$$
\begin{aligned}
\langle G(u^k) \ - \ G(z^{k,m}), u^k - z^{k,m}\rangle &\leq L\|z^{k,m} - u^k\|^2 \\
&\leq \ L\langle T_k(u^k, z^{k,m}), z^{k,m} - u^k\rangle/\tau' \\
&\leq \ (1-\alpha)(\tilde{\theta}\beta^m)^{-1}\langle T_k(u^k, z^{k,m}), z^{k,m} - u^k\rangle,
\end{aligned}
$$

when $(\tilde{\theta}\beta^m) \leq (1-\alpha)\tau'/L$ and part (ii) holds with $\theta' = \min\{\tilde{\theta}, (1-\alpha)\tau'/L\}$.
□

REMARK 1.1. Relations (1.77) and (1.68) imply

$$
\begin{aligned}
\langle G(u^k) - G(z^{k,m}), u^k - z^{k,m}\rangle &\leq (1-\alpha)(\tilde{\theta}\beta^m)^{-1}\langle T_k(u^k, z^{k,m}), z^{k,m} - u^k\rangle \\
&\leq (1-\alpha)\langle G(u^k), u^k - z^{k,m}\rangle,
\end{aligned}
$$

or equivalently,

$$
\langle G(z^{k,m}), z^{k,m} - u^k\rangle \leq \alpha\langle G(u^k), z^{k,m} - u^k\rangle,
$$

which coincides with (1.64). Hence, if we replace rule (1.64) by (1.77), then all the assertions of Section 1.3 concerning Method 1.2 remain valid. Similarly, we can replace rule (1.42) with the following:

$$\langle G(u^k + \beta^m \tilde{\theta} p^k) - G(u^k), p^k \rangle \leq (1 - \alpha)\langle T_k(u^k, z^k), p^k \rangle,$$

since, by (1.44),

$$\langle T_k(u^k, z^k), p^k \rangle \leq -\langle G(u^k), p^k \rangle.$$

On the other hand, the proofs of Lemmas 1.3.6 and 1.4.2 show that the methods admit the reverse substitution. Namely, we can use rule (1.64) instead of (1.77) in Method 1.3 as well.

Now, taking into account the properties of T_k and the definitions (1.78), (1.79), we conclude that, by using the argument similar to that in Theorem 1.3.5, the following convergence theorem is true.

Theorem 1.4.1. *Let a sequence $\{u^k\}$ be generated by Method 1.3. Then:*

(i) If the method terminates at the kth iteration, $v^k \in U^*$.

(ii) If $\{u^k\}$ is infinite, there exists a limit point u^* of $\{u^k\}$ which lies in U^*.

(iii) If $\{u^k\}$ is infinite, and (1.47) holds, we have

$$\lim_{k \to \infty} u^k = u^* \in U^*.$$

We now obtain rates of convergence for Method 1.3.

Theorem 1.4.2. *Let an infinite sequence $\{u^k\}$ be generated by Method 1.3, where $\{\gamma_k\}$ is chosen in accordance with rule (1.52).*

(i) It holds that

$$\liminf_{k \to \infty} \left[\langle T_k(u^k, v^k), v^k - u^k \rangle (k + 1) \right] = 0.$$

(ii) If (1.47) and Assumption (A5) hold, then $\{\|u^k - \pi_{U^*}(u^k)\|\}$ converges to zero in a linear rate.

Proof. Taking into account Lemmas 1.3.5 (ii), 1.4.1 (ii) and 1.4.2 (ii), the properties of T_k, and (1.78), (1.79), we have

$$
\begin{aligned}
(\sigma_k \|t^k\|)^2 &= \alpha^2 \langle T_k(u^k, v^k), v^k - u^k \rangle [\langle T_k(u^k, v^k), v^k - u^k \rangle / (\|t^k\| \theta_k)^2] \\
&\geq \alpha^2 \langle T_k(u^k, v^k), v^k - u^k \rangle \\
&\quad [\tau' \|u^k - v^k\|^2 / \|\theta_k [G(v^k) - G(u^k)] - T_k(u^k, v^k)\|)^2] \\
&\geq \alpha^2 \tau' \langle T_k(u^k, v^k), v^k - u^k \rangle / (\tilde{\theta} L + \tau'')^2 \\
&= \alpha' \langle T_k(u^k, v^k), v^k - u^k \rangle,
\end{aligned}
$$

where L is the corresponding Lipschitz constant. It follows that

$$(\sigma_k \|t^k\|)^2 \geq \alpha' \langle T_k(u^k, v^k), v^k - u^k \rangle \qquad (1.82)$$

where $\alpha' = \alpha^2 \tau' / (\tilde{\theta} L + \tau'')^2$. By using Lemma 1.4.1 (iii) and the argument similar to that in Theorem 1.3.2 we now conclude that assertion (i) is true. In case (ii), from (1.52), (1.82) and Lemma 1.4.1 (i) we have

$$\|u^{k+1} - \pi_{U^*}(u^{k+1})\|^2 \;\leq\; \|u^k - \pi_{U^*}(u^k)\|^2$$
$$-\gamma(2-\gamma)\alpha' \langle T_k(u^k, v^k), v^k - u^k \rangle; \quad (1.83)$$

besides, $0 < \theta' \leq \theta_k \leq \tilde{\theta}$ due to Lemma 1.4.2 (ii), hence Assumption (A5) is now applicable. Therefore, $\langle T_k(u^k, v^k), v^k - u^k \rangle \to 0$ as $k \to \infty$ and

$$\langle T_k(u^k, v^k), v^k - u^k \rangle \geq \mu \|u^k - \pi_{U^*}(u^k)\|^2 \qquad (1.84)$$

for k large enough. Combining (1.83) and (1.84), we conclude that assertion (ii) is also true. $\square$

Theorem 1.4.3. *Let a sequence $\{u^k\}$ be constructed by Method 1.3. Suppose that (1.47) and Assumption (A6) hold. Then the method terminates with a solution.*

Proof. Under the assumptions of the present theorem, (1.82) holds too. Lemma 1.4.1 (iii), (1.68) and (1.82) now imply

$$\liminf_{k \to \infty} \|u^k - v^k\| = 0,$$

i.e., (1.70) is fulfilled. The rest is proved by the same argument as in the proof of Theorem 1.9. $\square$

Thus, Method 1.3 finds a solution in a finite number of iterations under the same assumptions as those in Section 1.3. However, the estimates in Theorem 1.4.2 are better than the corresponding estimates in Theorems 1.3.2, 1.3.6, 1.3.3 (i) and 1.3.7 (i), respectively. On the other hand, Methods 1.1 and 1.2 need not just the projection operation in Step 2.

1.4.2 A Modified Rule for Computing the Descent Direction

We now describe a CR method which uses a different rule of computing the descent direction. This rule, in particular, allows one to simplify the numerical implementation of the main iteration. We consider the method under Hypothesis $(H1.3)$ of Section 1.3.

Method 1.4. *Step 0 (Initialization)*: Choose a point $u^0 \in V$, a sequence $\{\gamma_k\}$ satisfying (1.40), a family of mappings $\{T_k : V \times V \to R^n\}$ satisfying Assumptions (A1) – (A4), and choose a sequence of mappings $\{P_k\}$, where $P_k \in \mathcal{F}(V_k)$, $U \subseteq V_k \subseteq V$ for $k = 0, 1, \ldots$ Choose numbers $\alpha \in (0,1)$, $\beta \in (0,1)$, $\tilde{\theta} > 0$. Set $k := 0$.

   Step 1 (Auxiliary procedure D_k):

   Step 1.1: Find m as the smallest number in Z_+ such that (1.77) holds, where $z^{k,m}$ is a solution of the auxiliary problem of finding $\bar{z} \in U$ such that (1.65) holds.

   Step 1.2: Set $\theta_k := \beta^m \tilde{\theta}$, $v^k := z^{k,m}$. If $u^k = v^k$ or $G(v^k) = 0$, stop.

Step 2 (Main iteration): Set

$$g^k \; := \; G(v^k) - G(u^k) - \theta_k^{-1} T_k(u^k, v^k), \qquad (1.85)$$

$$\sigma_k \; := \; \langle g^k, u^k - v^k \rangle / \|g^k\|^2, \qquad (1.86)$$

$$u^{k+1} \; := \; P_k(u^k - \gamma_k \sigma_k g^k), \qquad (1.87)$$

$k := k + 1$ and go to Step 1.

In comparison with Method 1.3, we now set g^k to be equal t^k, but rule (1.86) for computing the stepsize σ_k coincides with those in Methods 1.1 and 1.2. The other essential feature of this method, unlike Methods 1.1 – 1.3, is that the sets V_k and V can be determined rather arbitrarily within $U \subseteq V_k \subseteq V \subseteq R^n$. Namely, one can choose different V_k at each step. Note that VI's with $U \subset V$ and V being defined by simple constraints arise in many applications (e.g. see Section 1 and Chapter 3), so that just the proposed method can rather simply handle such a situation. Note also that due to (1.87) $\{u^k\}$ need not be feasible if $U \subset V$.

We first obtain estimates for θ_k and $\|u^k - z^{k,m}\|$.

Lemma 1.4.3. *Fix $\tilde{u} \in U$. It holds that*

$$\tilde{\theta}\|G(u^k)\| + \tau_k''\|u^k - \tilde{u}\| \ge \tau_k'\|\tilde{u} - z^{k,m}\| \quad \forall m \in Z_+, \qquad (1.88)$$

$$\|u^k - z^{k,m}\| \le \lambda_k \quad \forall m \in Z_+, \qquad (1.89)$$

$$\theta_k \ge \min\{\beta(1-\alpha)\tau_k'/L_k, \tilde{\theta}\}; \qquad (1.90)$$

where

$$\lambda_k = \tilde{\theta}\|G(u^k)\|/\tau_k' + (1 + \tau_k''/\tau_k')\|u^k - \tilde{u}\|, \qquad (1.91)$$

L_k is the Lipschitz constant for G on the set $\{u \in V \mid \|u - u^k\| \le \lambda_k\}$.

Proof. Applying (1.65) with $u = \tilde{u}$, $\bar{z} = z^{k,m}$ gives

$$\langle G(u^k) + (\tilde{\theta}\beta^m)^{-1} T_k(u^k, z^{k,m}), \tilde{u} - z^{k,m} \rangle \ge 0.$$

By the properties of T_k we then have

$$\begin{aligned} \tilde{\theta}\|G(u^k)\|\|\tilde{u} - z^{k,m}\| \; &\ge \; (\tilde{\theta}\beta^m)\|G(u^k)\|\|\tilde{u} - z^{k,m}\| \\ &\ge \; (\tilde{\theta}\beta^m)\langle G(u^k), \tilde{u} - z^{k,m}\rangle \\ &\ge \; \langle T_k(u^k, z^{k,m}) - T_k(u^k, \tilde{u}), z^{k,m} - \tilde{u}\rangle \\ &+ \; \langle T_k(u^k, \tilde{u}) - T_k(u^k, u^k), z^{k,m} - \tilde{u}\rangle \\ &\ge \; \tau_k'\|z^{k,m} - \tilde{u}\|^2 - \tau_k''\|u^k - \tilde{u}\|\|z^{k,m} - \tilde{u}\|, \end{aligned}$$

i.e., (1.88) holds. It follows that

$$\begin{aligned} \|u^k - z^{k,m}\| \; &\le \; \|u^k - \tilde{u}\| + \|\tilde{u} - z^{k,m}\| \\ &\le \; \tilde{\theta}\|G(u^k)\|/\tau_k' + (\tau_k''/\tau_k')\|u^k - \tilde{u}\| + \|u^k - \tilde{u}\|, \end{aligned}$$

i.e., (1.89) with λ_k being defined in (1.91) also holds. From (1.89), the definition of L_k and the properties of T_k it follows that

$$\langle G(u^k) - G(z^{k,m}), u^k - z^{k,m} \rangle \leq L_k \|u^k - z^{k,m}\|^2$$
$$\leq L_k \langle T_k(u^k, z^{k,m}), z^{k,m} - u^k \rangle / \tau_k'$$
$$\leq (1-\alpha)(\tilde{\theta}\beta^m)^{-1}\langle T_k(u^k, z^{k,m}), z^{k,m} - u^k \rangle,$$

when $(1-\alpha)(\tilde{\theta}\beta^m)^{-1} \geq L_k/\tau_k'$. Taking into account (1.77) we obtain (1.90) and the proof is complete. $\qquad\square$

Next, by construction, the assertions of (i) – (iii) of Lemma 1.3.5 are true for Method 1.4 as well. Besides, we obtain a modification of (1.68).

Lemma 1.4.4. *It holds that*

$$\langle g^k, u^k - v^k \rangle \geq (\alpha/\theta_k)\langle T_k(u^k, v^k), v^k - u^k \rangle \geq (\alpha\tau_k'/\theta_k)\|u^k - v^k\|^2. \quad (1.92)$$

Proof. In case $u^k = v^k$ (1.92) obviously holds. Consider the case $u^k \neq v^k$. Then, by (1.77), (1.85), and the properties of T_k, we have

$$
\begin{aligned}
\langle g^k, u^k - v^k \rangle &= \langle G(v^k) - G(u^k) - \theta_k^{-1} T_k(u^k, v^k), u^k - v^k \rangle \\
&\geq [(1-\alpha)/\theta_k]\langle T_k(u^k, v^k), u^k - v^k \rangle \\
&\quad -(1/\theta_k)\langle T_k(u^k, v^k), u^k - v^k \rangle \\
&= (\alpha/\theta_k)\langle T_k(u^k, v^k), v^k - u^k \rangle \geq (\alpha\tau_k'/\theta_k)\|u^k - v^k\|^2,
\end{aligned}
$$

i.e., (1.92) holds. $\qquad\square$

Note that (1.92) implies $g^k \neq 0$ in case $u^k \neq v^k$, i.e., Step 2 of the method is then well defined.

Lemma 1.4.5. *It holds that:*
  (i)

$$\|u^{k+1} - u^*\|^2 \leq \|u^k - u^*\|^2 - \gamma_k(2 - \gamma_k)(\sigma_k\|g^k\|)^2 \qquad \forall u^* \in U^d.$$

  (ii) $\{u^k\}$ *is bounded.*
  (iii) $\sum\limits_{k=0}^{\infty} \gamma_k(2 - \gamma_k)(\sigma_k\|g^k\|)^2 < \infty.$
  (iv) For each limit point u^* of $\{u^k\}$ such that $u^* \in U^d$ we have

$$\lim_{k \to \infty} u^k = u^*.$$

  (v) $\{v^k\}$ *is bounded.*
  (vi) $\theta_k \geq \theta' > 0$ *for* $k = 0, 1, \ldots$

Proof. Take any $u^* \in U^d$. Then

$$\langle G(v^k), v^k - u^* \rangle \geq 0,$$

whereas applying (1.65) with $u = u^*$, $\bar{z} = z^{k,m} = v^k$ yields

$$\langle -G(u^k) - \theta_k^{-1} T_k(u^k, v^k), v^k - u^* \rangle \geq 0.$$

Adding both the inequalities above gives $\langle g^k, v^k - u^* \rangle \geq 0$. It follows that

$$\begin{aligned} \langle g^k, u^k - u^* \rangle &= \langle g^k, u^k - v^k \rangle + \langle g^k, v^k - u^* \rangle \\ &\geq \sigma_k \|g^k\|^2 \geq 0, \end{aligned}$$

i.e., (1.36) holds. Following the proof of Lemma 1.2.3 and using V_k instead of U, we now see that part (i) holds. Therefore, all the assertions of Lemma 1.2.4 are true for Method 1.4. This proves parts (ii) – (iv). Next, since $\{u^k\}$ is bounded, (1.89), (1.91) and (1.46) imply the boundedness of $\{v^k\}$, i.e., part (v) holds too. Moreover, we then have $\lambda_k \leq \lambda' < \infty$ and $L_k \leq L' < \infty$ for $k = 0, 1, \dots$ By (1.90), we now get (vi). □

Now, taking into account (1.92) and the properties of T_k, we conclude that by using the argument similar to that in Theorem 1.3.5, the following convergence theorem is true.

Theorem 1.4.4. *Let a sequence $\{u^k\}$ be generated by Method 1.4. Then:*
(i) If the method terminates at the kth iteration, $v^k \in U^*$.
(ii) If $\{u^k\}$ is infinite, there exists a limit point u^* of $\{u^k\}$ which lies in U^*.
(iii) If $\{u^k\}$ is infinite, and (1.47) holds, we have

$$\lim_{k \to \infty} u^k = u^* \in U^*.$$

We are now ready to obtain rates of convergence for Method 1.4.

Theorem 1.4.5. *Let an infinite sequence $\{u^k\}$ be generated by Method 1.4, where $\{\gamma_k\}$ is chosen in accordance with rule (1.52).*
(i) It holds that

$$\liminf_{k \to \infty} \left[\langle T_k(u^k, v^k), v^k - u^k \rangle (k+1) \right] = 0.$$

(ii) If (1.47) and Assumption (A5) hold, then $\{\|u^k - \pi_{U^*}(u^k)\|\}$ converges to zero in a linear rate.

Proof. Taking into account Lemmas 1.4.4, 1.4.5, the properties of T_k and (1.85), (1.86), we have

$$
\begin{aligned}
(\sigma_k \|g^k\|)^2 &= (\langle g^k, u^k - v^k \rangle / \|G(v^k) - G(u^k) - \theta_k^{-1} T_k(u^k, v^k)\|)^2 \\
&\geq \alpha^2 \langle T_k(u^k, v^k), v^k - u^k \rangle \\
&\quad \tau' \|u^k - v^k\|^2 / \|\theta_k [G(v^k) - G(u^k)] - T_k(u^k, v^k)\|^2 \\
&\geq \alpha^2 \langle T_k(u^k, v^k), v^k - u^k \rangle \\
&\quad \tau' \|u^k - v^k\|^2 / (\theta_k L \|u^k - v^k\| + \tau'' \|u^k - v^k\|)^2 \\
&\geq \alpha^2 \tau' \langle T_k(u^k, v^k), v^k - u^k \rangle / (\bar{\theta} L + \tau'')^2 \\
&= \alpha' \langle T_k(u^k, v^k), v^k - u^k \rangle,
\end{aligned}
$$

where L is the corresponding Lipschitz constant. It follows that

$$
(\sigma_k \|g^k\|)^2 \geq \alpha' \langle T_k(u^k, v^k), v^k - u^k \rangle \tag{1.93}
$$

where $\alpha' = \alpha^2 \tau' / (\bar{\theta} L + \tau'')^2$. By using Lemma 1.4.5 (iii) and the argument similar to that in Theorem 1.3.2 we now conclude that assertion (i) is true. Assertion (ii) is proved by the same argument as in the proof of Theorem 1.4.2 with replacing (1.82) by (1.93). □

Similarly, following the proof of Theorem 1.4.3 and replacing (1.82) by (1.93), we obtain the following condition of finiteness for Method 1.4.

Theorem 1.4.6. *Let a sequence $\{u^k\}$ be constructed by Method 1.4. Suppose that (1.47) and Assumption (A6) hold. Then the method terminates with a solution.*

Thus, Methods 1.3 and 1.4 attain the same rates of convergence, although they use different rules for computing the stepsize and descent direction in the main process.

1.5 CR Method Based on a Frank-Wolfe Type Auxiliary Procedure

Basic Scheme I admits wide variety of auxiliary procedures which provide the key property (1.36). In fact, most relaxation methods can be regarded as a basis for constructing such an auxiliary procedure. In this section, in addition to the methods of Sections 1.3 and 1.4, we construct a CR method whose auxiliary procedure is based on an iteration of the Frank-Wolfe type method. The description of the Frank-Wolfe method can be found in [58] and, also, in [136, 42, 183]. In this section, we suppose that Hypothesis $(H1.3)$ of Section 1.3 is fulfilled and, also, that the feasible set U is bounded.

1.5.1 Description of the Method

We describe the CR method with a Frank-Wolfe type auxiliary procedure as follows.

Method 1.5. *Step 0 (Initialization)*: Choose a point $u^0 \in U$, a sequence $\{\gamma_k\}$ satisfying (1.40), and a sequence of mappings P_k, where $P_k \in \mathcal{F}(U)$ for $k = 0, 1, \ldots$ Choose numbers $\alpha \in (0,1)$, $\beta \in (0,1)$, $\tilde{\theta} \in (0,1]$. Set $k := 0$.

Step 1 (Auxiliary procedure D_k):

Step 1.1: Solve the auxiliary variational inequality problem of finding $z^k \in U$ such that

$$\langle G(u^k), v - z^k \rangle \geq 0 \quad \forall v \in U \tag{1.94}$$

and set $p^k := z^k - u^k$. If $p^k = 0$, stop.

Step 1.2: Determine m as the smallest number in Z_+ such that

$$\langle G(u^k + \beta^m \tilde{\theta} p^k), p^k \rangle \leq \alpha \langle G(u^k), p^k \rangle, \tag{1.95}$$

set $\theta_k := \beta^m \tilde{\theta}$, $v^k := u^k + \theta_k p^k$. If $G(v^k) = 0$, stop.

Step 2 (Main iteration): Set

$$g^k := G(v^k), \sigma_k := \langle G(v^k), u^k - v^k \rangle / \|g^k\|^2, u^{k+1} := P_k(u^k - \gamma_k \sigma_k g^k), \tag{1.96}$$

$k := k + 1$ and go to Step 1.

Note that the auxiliary problem (1.94) coincides with (1.41) in the case where $T_k \equiv 0$. In other words, the auxiliary problem in Method 1.5 can be viewed as a "degenerate" version of that in Method 1.1. We insert the additional assumption of boundedness in order to guarantee for the problem (1.94) to have a solution. Besides, the problem (1.94) need not have a unique solution. Nevertheless, the problem (1.94) has certain advantages over (1.41). For instance, if U is a polyhedral set, (1.94) corresponds to a linear programming problem and its solution becomes much simpler for computation in comparison with that of problem (1.41) which then corresponds to an LCP.

It is clear that Method 1.5 falls into Basic Scheme I, however, its properties are different from those of Method 1.1. In fact, we have the following analogue of Lemma 1.3.1.

Lemma 1.5.1. *(i) Problem (1.94) is solvable.*
(ii) If $\langle G(u^k), u^k - z^k \rangle = 0$, then $u^k \in U^*$.

Nevertheless, the proof of its convergence, in general, can be argued similarly.

1.5.2 Convergence

We first give an analogue of Lemmas 1.3.2 and 1.3.3 for Method 1.5.

Lemma 1.5.2. *It holds that:*
(i)

$$\|u^{k+1} - u^*\|^2 \leq \|u^k - u^*\|^2 - \gamma_k(2 - \gamma_k)(\sigma_k\|g^k\|)^2 \qquad \forall u^* \in U^d.$$

(ii) $\sum_{k=0}^{\infty} \gamma_k(2 - \gamma_k)(\sigma_k\|g^k\|)^2 < \infty.$
(iii) For each limit point u^* of $\{u^k\}$ such that $u^* \in U^d$, we have

$$\lim_{k \to \infty} u^k = u^*.$$

(iv) $\theta_k \geq \alpha'\langle G(u^k), p^k\rangle$ for $k = 0, 1, \ldots,$ where $\alpha' > 0.$

Proof. From (1.94) and (1.95) it follows that

$$\begin{aligned}
\sigma_k\|g^k\|^2 &= \theta_k\langle G(v^k), u^k - z^k\rangle \\
&\geq \alpha\theta_k\langle G(u^k), u^k - z^k\rangle \geq 0.
\end{aligned}$$

For each $u^* \in U^d$ we then have

$$\begin{aligned}
\langle g^k, u^k - u^*\rangle &= \langle G(v^k), u^k - v^k\rangle + \langle G(v^k), v^k - u^*\rangle \\
&\geq \sigma_k\|g^k\|^2 \geq 0,
\end{aligned}$$

i.e., (1.36) holds, and assertions (i) - (iii) follow from Lemmas 1.2.3 and 1.2.4. Next, for each $\theta \in (0, 1]$,

$$\begin{aligned}
\langle G(u^k + \theta p^k), p^k\rangle &= \langle G(u^k), p^k\rangle + \langle G(u^k + \theta p^k) - G(u^k), p^k\rangle \\
&\leq \langle G(u^k), p^k\rangle + L\theta\|p^k\|^2,
\end{aligned}$$

where L is the Lipschitz constant for G on U. Hence, if

$$\theta \leq -(1 - \alpha)\langle G(u^k), p^k\rangle/(Ld^2),$$

where

$$d = \sup_{a,b \in U} \|a - b\| < \infty,$$

we have

$$\langle G(u^k + \theta p^k), p^k \rangle \leq \langle G(u^k), p^k \rangle - (1 - \alpha)\langle G(u^k), p^k \rangle \|p^k\|^2 / \|d\|^2$$
$$\leq \alpha \langle G(u^k), p^k \rangle$$

and we see that $\theta_k \geq \min\{\bar{\theta}, -(1 - \alpha)\langle G(u^k), p^k \rangle / (Ld^2)\}$, hence part (iv) holds, too.   □

We are now in a position to establish a convergence result for Method 1.5.

Theorem 1.5.1. *Let a sequence $\{u^k\}$ be generated by Method 1.5. Then:*

(i) If the method terminates at Step 1.1 (Step 1.2) of the kth iteration, $u^k \in U^*$ ($v^k \in U^*$).

(ii) If $\{u^k\}$ is infinite, there exists a limit point u^* of $\{u^k\}$ which lies in U^*.

(iii) If $\{u^k\}$ is infinite and (1.47) holds, we have

$$\lim_{k \to \infty} u^k = u^* \in U^*.$$

Proof. Part (i) follows directly from the stopping rule and Lemma 1.5.1 (ii). Next, if the sequence $\{u^k\}$ is infinite, from (1.95), (1.96) and Lemma 1.5.2 (iv) it follows that

$$\sigma_k \|g^k\| = \langle G(v^k), u^k - v^k \rangle / \|G(v^k)\| \geq \alpha\alpha' \langle G(u^k), u^k - z^k \rangle^2 / L.$$

Since $\{u^k\}$ and $\{z^k\}$ are contained in the bounded set U, they have limit points. Without loss of generality we suppose that

$$\lim_{k \to \infty} u^k = u^*, \lim_{k \to \infty} z^k = z^*. \tag{1.97}$$

Now, from Lemma 1.5.2 (ii) and (1.40) we obtain

$$\liminf_{k \to \infty} \langle G(u^k), u^k - z^k \rangle = \langle G(u^*), u^* - z^* \rangle = 0.$$

If we suppose that $u^* \notin U^*$, then on account of Lemma 1.5.1 (ii) we have

$$\langle G(u^*), u^* - z(u^*) \rangle > 0, \tag{1.98}$$

where $z(u^*)$ is a solution to (1.94) with $u^k = u^*$, i.e.,

$$\langle G(u^*), v - z(u^*) \rangle \geq 0 \quad \forall v \in U.$$

On the other hand, from (1.94), (1.97) we obtain

$$\langle G(u^*), v - z^* \rangle \geq 0 \quad \forall v \in U,$$

hence, by (1.98),

$$0 < \langle G(u^*), u^* - z(u^*) \rangle = \langle G(u^*), u^* - z^* \rangle + \langle G(u^*), z^* - z(u^*) \rangle$$
$$\leq \langle G(u^*), u^* - z^* \rangle,$$

which is a contradiction. Part (iii) now follows directly from Lemma 1.5.2 (iii).   □

1.6 CR Method for Variational Inequalities with Nonlinear Constraints

In this section, we consider the case of VI (1.1) subject to nonlinear constraints. More precisely, we suppose that Hypothesis $(H1.3)$ of Section 1.3 is fulfilled, with U having the form

$$U = \{u \in R^n \mid h_i(u) \leq 0 \quad i = 1, \ldots, m\},$$

where $h_i : R^n \to R$, $i = 1, \ldots, m$ are continuously differentiable and convex functions, there exists a point $\bar{u}$ such that $h_i(\bar{u}) < 0$ for $i = 1, \ldots, m$. Since the functions h_i, $i = 1, \ldots, m$ need not be affine, the auxiliary problems at Step 1 of the methods of the previous sections cannot in general be solved by finite algorithms. Therefore, we need a finite auxiliary procedure for this case. Such a CR method is constructed in this section.

1.6.1 A Modified Basic Scheme for CR Methods

All the methods from the previous sections follow Basic Scheme I from Section 1.2. In order to construct CR methods in the nonlinear constrained case, we need the other basic scheme. In fact, each auxiliary procedure at Step 1 of Methods 1.1 – 1.5 either generates a descent direction g^k (serious step), or terminates with a solution. The new basic scheme has to involve the situation of a null step, where the auxiliary procedure yields the zero vector, but the current iterate is not a solution of VI (1.1). The null step usually occurs if the current tolerances are too large, hence they must diminish. The corresponding basic scheme can be described as follows.

Basic Scheme II. Choose a point $u^0 \in U$ and a sequence $\{\gamma_j\}$ such that

$$\gamma_j \in [0, 2], j = 0, 1, \ldots; \quad \sum_{j=0}^{\infty} \gamma_j(2 - \gamma_j) = \infty. \qquad (1.99)$$

Also, choose a sequence of mappings $\{P_k\}$, where $P_k \in \mathcal{F}(U)$ for $k = 0, 1, \ldots$ Set $\lambda(0) := 1$, $j(0) := 0$.

At the kth iteration, $k = 0, 1, \ldots$, we have a point $u^k \in U$ and numbers $\lambda(k)$, $j(k)$.

Step 1: Apply Procedure $\tilde{D}_k$ and obtain the output: a point v^k and a number $\lambda(k + 1)$. If $G(v^k) = 0$, stop.

Step 2: If $\lambda(k + 1) > \lambda(k)$, set $y^{\lambda(k)} := u^{k+1} := u^k$, $j(k + 1) := j(k)$. Otherwise, set $j(k + 1) := j(k) + 1$,

$$g^k := G(v^k), \sigma_k := \langle g^k, u^k - v^k \rangle / \|g^k\|^2, u^{k+1} := P_k(u^k - \gamma_{j(k)}\sigma_k g^k).$$

The iteration has been completed.

According to the description, in the case of a null step, we increase $\lambda(\cdot)$, in the case of a descent (serious) step, we increase $j(\cdot)$. Thus, $\lambda(k)$ (respectively, $j(k)$) is a counter for null (respectively, serious) steps. The input of the auxiliary procedure $\tilde{D}_k$ consists of the current iterate u^k and the current value of $\lambda(\cdot)$, the output contains an auxiliary point v^k and the new value of $\lambda(\cdot)$. In general, Basic Scheme II follows the process (1.35), (1.36) as well as Basic Scheme I. In particular, in the case of a descent step we must have

$$\langle G(v^k), u^k - v^k \rangle > 0,$$

which leads to (1.36).

The key properties of Basic Scheme II can be stated as follows.

Lemma 1.6.1. *Suppose that a sequence $\{u^k\}$ be constructed by Basic Scheme II and that (1.36) holds. Then:*
(i)

$$\|u^{k+1} - u^*\|^2 \le \|u^k - u^*\|^2 - \gamma_{j(k)}(2 - \gamma_{j(k)})(\sigma_k\|g^k\|)^2 \qquad \forall u^* \in U^d.$$

(ii)

$$\sum_{k=0}^{\infty} \gamma_{j(k)}(2 - \gamma_{j(k)})(\sigma_k\|g^k\|)^2 < \infty.$$

(iii) If $\sigma_k\|g^k\| \ge \tilde{\sigma} > 0$ as $\lambda(k) \le \bar{l} < \infty$, the sequence $\{y^l\}$ is infinite.
(iv) For each limit point u^* of $\{u^k\}$ such that $u^* \in U^d$ we have

$$\lim_{k \to \infty} u^k = u^*.$$

Proof. Part (i) is argued similarly to the proof of Lemma 1.2.3. Parts (ii) and (iv) follow directly from (i). Next, in case (iii), assume, for contradiction, that $\{y^l\}$ is finite. Then $\lambda(k) \le \bar{l} < \infty$, i.e., we have $\sigma_k\|g^k\| \ge \tilde{\sigma} > 0$. It follows that

$$\sum_{k=0}^{\infty}(\sigma_k\|g^k\|)^2 = \infty,$$

which contradicts part (ii) due to (1.99). $\qquad\qquad\qquad\qquad\qquad\qquad\quad$ $\square$

1.6.2 Description of the Method

The method for solving VI (1.1) under the assumptions above can be described as follows. Set

$$I_\varepsilon(u) = \{i \mid 1 \le i \le m, \quad h_i(u) \ge -\varepsilon\}.$$

Method 1.6. *Step 0 (Initialization):* Choose a point $u^0 \in U$, a sequence $\{\gamma_j\}$ satisfying (1.99), sequences $\{\varepsilon_l\}$ and $\{\eta_l\}$ such that

$$\{\varepsilon_l\} \searrow 0, \{\eta_l\} \searrow 0. \tag{1.100}$$

Also, choose a sequence of mappings $\{P_k\}$, where $P_k \in \mathcal{F}(U)$ for $k = 0, 1, \ldots$
Choose numbers $\alpha \in (0,1)$, $\beta \in (0,1)$, $\theta \in (0,1]$, and $\mu_i > 0$ for $i = 1, \ldots, m$.
Set $\lambda(0) := 1$, $j(0) := 0$, $k := 0$.

Step 1 (Auxiliary procedure $\tilde{D}_k$): Set $l := \lambda(k)$.

Step 1.1 : Find the solution $(\tau_{k,l}, p^{k,l})$ of the problem

$$\min \quad \to \quad \tau \tag{1.101}$$

subject to

$$
\begin{aligned}
\langle G(u^k), p \rangle &\leq \tau, \\
\langle \nabla h_i(u^k), p \rangle &\leq \mu_i \tau \qquad i \in I_{\varepsilon_l}(u^k), \\
|p_j| &\leq 1 \quad j = 1, \ldots, n.
\end{aligned}
\tag{1.102}
$$

Step 1.2: If $\tau_{k,l} \geq -\eta_l$, set $v^k := u^k$, $\lambda(k+1) := l+1$, $j(k+1) := j(k)$ and go to Step 2. (*null step*)

Step 1.3: Determine m as the smallest number in Z_+ such that

$$u^k + \beta^m \tilde{\theta} p^{k,l} \in U, \langle G(u^k + \beta^m \tilde{\theta} p^{k,l}), p^{k,l} \rangle \leq \alpha \langle G(u^k), p^{k,l} \rangle, \tag{1.103}$$

set $\theta_k := \beta^m \tilde{\theta}$, $v^k := u^k + \theta_k p^{k,l}$, $\lambda(k+1) := \lambda(k)$. (*descent step*)

Step 2 (Main iteration): If $\lambda(k+1) > \lambda(k)$, set $y^{\lambda(k)} := u^{k+1} := u^k$, $j(k+1) := j(k)$, $g^k := 0$, $\sigma_k := 0$, $k := k+1$ and go to Step 1. Otherwise, set

$$
\begin{aligned}
g^k &:= G(v^k), \sigma_k := \langle G(v^k), u^k - v^k \rangle / \|g^k\|^2, \\
u^{k+1} &:= P_k(u^k - \gamma_{j(k)} \sigma_k g^k),
\end{aligned}
\tag{1.104}
$$

$j(k+1) := j(k) + 1$, $k := k+1$ and go to Step 1.

It is easy to see that the auxiliary procedure in Step 1 is an analogue of an iteration of the feasible direction method [225]. Deriving its properties, we follow the methodology of [225, 183]. Denote by $D(u^k, \varepsilon_l)$ the set of points $(\tau, p) \in R^{n+1}$ satisfying (1.102).

Lemma 1.6.2. *(i)* $D(u^k, \varepsilon_l) \neq \emptyset$.

(ii) The solution of the problem (1.101), (1.102) always exists.

(iii) A point $u^k \in U$ is a solution of VI (1.1) if and only if $\varepsilon_l = 0$ and $\tau_{k,l} = 0$ in (1.101), (1.102).

(iv) If $\theta_k > 0$, $g^k \neq 0$.

Proof. First we have $D(u^k, \varepsilon_l) \neq \emptyset$ since $0 \in D(u^k, \varepsilon_l)$. Part (ii) follows from (i) and from the fact that τ is bounded from below on $D(u^k, \varepsilon_l)$. Next, if $(0, p^{k,l})$ is a solution of (1.101), (1.102) with $\varepsilon_l = 0$, then

$$\langle G(u^k), p \rangle \geq 0$$

for each $p \in R^n$ such that $\langle \nabla h_i(u^k), p \rangle \leq 0$ for $i \in I_0(u^k)$. Applying the well-known Farkas lemma (e.g. [225, Section 2.2, Theorem 1]) we see that there are nonnegative numbers y_i, $i = 1, \ldots, m$ such that

$$G(u^k) + \sum_{i=1}^{m} y_i \nabla h_i(u^k) = 0; \quad y_i \nabla h_i(u^k) = 0 \quad i = 1, \ldots, m;$$

hence (1.1), (1.19) (or (1.20)) with using R^l, G and h_i instead of R_+^l, F and f_i hold. Now, from Proposition 1.1.11 we obtain the desired result of (iii). In case (iv), if $\theta_k > 0$, then, taking into account (1.102) – (1.104), we have

$$0 > -\alpha \eta_l \theta_k > \alpha \tau_{k,l} \theta_k \geq \alpha \theta_k \langle G(u^k), p^{k,l} \rangle \geq \theta_k \langle G(v^k), p^{k,l} \rangle,$$

i.e. $G(v^k) \neq 0$. This proves part (iv). $\square$

From Lemma 1.6.2 (i), (ii) and (iv) it follows that Method 1.6 is well defined, if $\theta_k > 0$ for descent steps.

1.6.3 Convergence

In this section, we additionally suppose that the gradients $\nabla h_i : R^n \to R^n$, $i = 1, \ldots, m$ are locally Lipschitz continuous.

Lemma 1.6.3. (i) If $\tau_{k,l} < -\eta_l$, $\theta_k > 0$ for $k = 0, 1, \ldots$
   (ii) If $\{u^k\}$ is bounded, $\varepsilon_{\lambda(k)} \geq \varepsilon' > 0$, $\eta_{\lambda(k)} \geq \eta' > 0$ and $\tau_{k,\lambda(k)} < -\eta_{\lambda(k)}$, then $\theta_k \geq \theta' > 0$ for $k = 0, 1, \ldots$

Proof. Set $\bar{\mu} = \min_{i=1,\ldots,m} \mu_i$. First we show that $p^{k,l}$ is a feasible direction at u^k. Note that the last relation in (1.102) implies

$$\|p^{k,l}\|^2 \leq n. \tag{1.105}$$

For any $\theta \in (0, 1]$ and for some $\xi \in (0, 1)$, we have

$$h_i(u^k + \theta p^{k,l}) = h_i(u^k) + \langle \nabla h_i(u^k + \xi \theta p^{k,l}), p^{k,l} \rangle$$

for $i = 1, \ldots, m$. In case $i \in I_{\varepsilon_l}(u^k)$, we get

$$\begin{aligned}
h_i(u^k + \theta p^{k,l}) &\leq \theta(\langle \nabla h_i(u^k), p^{k,l} \rangle + L_{i,k} \theta \|p^{k,l}\|^2) \\
&\leq \theta(\mu_i \tau_{k,l} + L_{i,k} \theta \|p^{k,l}\|^2) \\
&\leq \theta(-\bar{\mu} \eta_l + L_{i,k} \theta n),
\end{aligned}$$

where $L_{i,k}$ is the corresponding Lipschitz constant for ∇h_i on $B(u^k, \sqrt{n})$. So, if we choose $\theta \leq \bar{\mu} \eta_l / (L_{i,k} n)$, then

$$h_i(u^k + \theta p^{k,l}) \leq 0. \tag{1.106}$$

Similarly, in the case of $i \notin I_{\varepsilon_l}(u^k)$ we have

$$h_i(u^k + \theta p^{k,l}) \leq -\varepsilon_l + \theta C_{i,k}\|p^{k,l}\| \leq -\varepsilon_l + \theta C_{i,k}\sqrt{n},$$

where $C_{i,k} \geq \|\nabla h_i(u^k + \xi\theta p^{k,l})\|$, so that (1.106) holds if $\theta \leq \varepsilon_l/(C_{i,k}\sqrt{n})$. Next, taking into account (1.102) and (1.105), we have

$$
\begin{aligned}
\langle G(u^k + \theta p^{k,l}), p^{k,l}\rangle &= \langle G(u^k), p^{k,l}\rangle + \langle G(u^k + \theta p^{k,l}) - G(u^k), p^{k,l}\rangle \\
&\leq \langle G(u^k), p^{k,l}\rangle + \theta\|p^{k,l}\|^2 L_k \\
&= \alpha\langle G(u^k), p^{k,l}\rangle + [(1-\alpha)\langle G(u^k), p^{k,l}\rangle + \theta\|p^{k,l}\|^2 L_k] \\
&\leq \alpha\langle G(u^k), p^{k,l}\rangle - [(1-\alpha)\eta_l - \theta L_k n]
\end{aligned}
$$

where L_k is the corresponding Lipschitz constant for G on $B(u^k, \sqrt{n})$. So, if we choose $\theta \leq (1-\alpha)\eta_l/(L_k n)$, then

$$\langle G(u^k + \theta p^{k,l}), p^{k,l}\rangle \leq \alpha\langle G(u^k), p^{k,l}\rangle.$$

Thus, $\theta_k > 0$ in any case and part (i) is true. In case (ii), we see that $L_k, L_{i,k} \leq L < \infty$ and $C_{i,k} \leq C < \infty$, so that $\theta_k \geq \theta' > 0$ where

$$\theta' = \min\{1, \bar{\mu}\eta'/(Ln), \varepsilon'/(C\sqrt{n}), (1-\alpha)\eta'/(Ln)\}.$$

The proof is complete.   □

Thus, rule (1.104) is also well defined. Now we obtain properties of Method 1.6 being based on Lemma 1.6.1.

Lemma 1.6.4. *Suppose* $\{u^k\}$ *is generated by Method 1.6. Then:*
(i)

$$\|u^{k+1} - u^*\|^2 \leq \|u^k - u^*\|^2 - \gamma_{j(k)}(2 - \gamma_{j(k)})(\sigma_k\|g^k\|)^2 \quad \forall u^* \in U_d^*.$$

(ii) $\{y^l\}$ *is infinite.*
(iii) $\{u^k\}$ *is bounded.*
(iv) For each limit point u^* *of* $\{u^k\}$ *such that* $u^* \in U^d$ *we have*

$$\lim_{k \to \infty} u^k = u^*.$$

Proof. In case (i) take any $u^* \in U^d$. The case $\lambda(k+1) > \lambda(k)$ is then trivial. Let $\lambda(k+1) = \lambda(k)$. Then, using (1.102) – (1.104), we have

$$
\begin{aligned}
\langle g^k, u^k - u^*\rangle &= \langle G(v^k), u^k - v^k\rangle + \langle G(v^k), v^k - u^*\rangle \\
&\geq \langle G(v^k), u^k - v^k\rangle = -\theta_k\langle G(v^k), p^{k,l}\rangle \\
&\geq -\alpha\theta_k\langle G(u^k), p^{k,l}\rangle \geq -\alpha\theta_k\tau_{k,l} \geq \alpha\theta_k\eta_l,
\end{aligned}
$$

hence (1.36) holds and part (i) follows from Lemma 1.6.1 (i). Analogously, we obtain part (iv) from Lemma 1.6.1 (iv). Part (iii) follows directly from (i).

In case (ii), assume for contradiction that $\{y^l\}$ is finite. Then $\lambda(k) \leq \bar{l} < \infty$ and

$$\sigma_k \|g^k\| = \langle G(v^k), u^k - v^k \rangle / \|G(v^k)\| \geq \alpha \theta_k \eta_l / \|G(v^k)\|.$$

Since $\{u^k\}$ is bounded, so are $\{v^k\}$ and $\{G(v^k)\}$ due to (1.105). On the other hand, $\eta_l \geq \eta' > 0$ and $\theta_k \geq \theta' > 0$ due to Lemma 1.6.3 (ii). It follows that

$$\sigma_k \|g^k\| \geq \alpha \theta' \eta' / C \geq \tilde{\sigma} > 0,$$

which contradicts Lemma 1.6.1 (iii). The proof is complete.   □

We are now in a position to establish a convergence result for Method 1.6.

Theorem 1.6.1. *Let a sequence $\{u^k\}$ be generated by Method 1.6. Then:*
(i) There exists a limit point u^* of $\{u^k\}$ which lies in U^*.
(ii) If (1.47) holds, we have

$$\lim_{k \to \infty} u^k = u^* \in U^*.$$

Proof. According to Lemma 1.6.4 (ii) and (iii), $\{y^l\}$ is infinite and bounded, so that it has a limit point $u^* \in U$. Without loss of generality we suppose $\lim_{l \to \infty} y^l = u^*$. Assume for contradiction that $u^* \notin U^*$. In this case, by Lemma 1.6.2 (iii), for the solution (τ^*, p^*) of the problem

$$\min \to \tau$$

subject to

$$
\begin{aligned}
\langle G(u^*), p \rangle &\leq \tau, \\
\langle \nabla h_i(u^*), p \rangle &\leq \mu_i \tau \quad i \in I_\varepsilon(u^*), \\
|p_j| &\leq 1 \quad j = 1, \ldots, n;
\end{aligned}
$$

we have $\tau^* < 0$ if $\varepsilon = 0$. But, by the continuity of h_i, there exists a number $\bar{\varepsilon} > 0$ such that $I_0(u^*) = I_{\bar{\varepsilon}}(u^*)$, hence $\tau^* < 0$ for any $\varepsilon \in [0, \bar{\varepsilon}]$ as well. On the one hand, by the continuity of G and ∇h_i, there exists a number t_1 such that

$$
\begin{aligned}
\langle G(y^l), p^* \rangle &\leq \tau^*/2, \\
\langle \nabla h_i(y^l), p^* \rangle &\leq \mu_i \tau^*/2 \quad i \in I_{\bar{\varepsilon}}(u^*), \\
|p_j^*| &\leq 1 \quad j = 1, \ldots, n;
\end{aligned}
$$

if $l > t_1$. On the other hand, by (1.100) and the continuity of h_i, there exists a number t_2 such that $\varepsilon_l \leq \bar{\varepsilon}$ and

$$I_{\varepsilon_l}(y^l) \subseteq I_{\bar{\varepsilon}}(y^l) \subseteq I_{\bar{\varepsilon}}(u^*)$$

if $l > t_2$. Therefore, if $l > \bar{t} = \max\{t_1, t_2\}$, then $(\tau^*/2, p^*) \in D(y^l, \varepsilon_l)$. We denote by k_l an iteration number such that $u^{k_l} = y^l$. Then $-\tau^*/2 \leq -\tau_{k_l, l} \leq$

η_l for l large enough. But this contradicts (1.100). So, the assertion of (i) is true. Part (ii) follows from Lemma 1.6.4 (iv).   □

Note that the auxiliary problem (1.101), (1.102) is a linear programming problem, so that its solution can be found by the well-known algorithms in a finite number of iterations. On the other hand, in order to implement a mapping $P_k \in \mathcal{F}(U)$ we can use finite procedures from Section 4.1. Thus, Method 1.6 is implementable completely under the above assumptions.

2. Variational Inequalities with Multivalued Mappings

In this chapter, we consider combined relaxation (CR) methods for solving variational inequalities which involve a multivalued mapping or a nonsmooth function.

2.1 Problem Formulation and Basic Facts

In this section, we give some facts from the theory of generalized variational inequality problems and their relations to other problems of Nonlinear Analysis. Thus, the results in this section can be viewed as extensions of those in Section 1.1.

2.1.1 Existence and Uniqueness Results

Let U be a nonempty, closed and convex subset of the n-dimensional Euclidean space R^n, $G : U \to \Pi(R^n)$ a multivalued mapping. The *generalized variational inequality problem* (GVI for short) is the problem of finding a point $u^* \in U$ such that

$$\exists g^* \in G(u^*), \quad \langle g^*, u - u^* \rangle \geq 0 \quad \forall u \in U. \tag{2.1}$$

In order to obtain existence results for GVI, we need several continuity properties of multivalued mappings.

Definition 2.1.1. [15, 159, 14, 127] Let W and V be convex sets in R^n, $W \subseteq V$, and let $Q : V \to \Pi(R^n)$ be a multivalued mapping. The mapping Q is said to be

(a) *upper semicontinuous (u.s.c.)* on W, if for each point $v \in W$ and for each open set Z such that $Z \supseteq G(v)$, there is a heighborhood X of v such that $Z \in G(w)$ whenever $w \in X \cup W$;

(b) *closed* on W, if for each pair of sequences $\{u^k\} \to u$, $\{q^k\} \to q$ such that $u^k \in W$ and $q^k \in Q(u^k)$, we have $q \in Q(u)$.

(c) a *K-mapping* on W, if it is u.s.c. on W and has nonempty convex and compact values;

(d) *u-hemicontinuous* on W, if for all $u \in W$, $v \in W$ and $\alpha \in [0, 1]$, the mapping $\alpha \mapsto \langle T(u + \alpha w), w \rangle$ with $w = v - u$ is u.s.c. at 0^+.

The following proposition gives the relationships between u.s.c. and closed mappings.

Proposition 2.1.1. *[159, Chapter 1, Lemma 4.4] Suppose $T : V \to \Pi(R^n)$ is a multivalued mapping, W is a convex subset of V.*

(i) If T is u.s.c. on W and has closed values, then it is closed on W.

(ii) If T is closed and for each compact set $X \subseteq W$, the set $T(X)$ is compact, then T is u.s.c. on W.

Proposition 2.1.2. *[53, 6, 70] Let $G : U \to \Pi(R^n)$ be a K-mapping. Suppose at least one of the following assumptions holds:*

(a) The set U is bounded;

(b) there exists a nonempty bounded subset W of U such that for every $u \in U\backslash W$ there is $v \in W$ with

$$\langle g, u - v \rangle > 0 \quad \forall g \in G(u).$$

Then GVI (2.1) has a solution.

The solution of GVI (2.1) is closely related with that of the corresponding *dual generalized variational inequality problem* (DGVI for short), which is to find a point $u^* \in U$ such that

$$\forall \, u \in U \text{ and } \forall g \in G(u) : \langle g, u - u^* \rangle \geq 0. \tag{2.2}$$

We denote by U^* (respectively, by U^d) the solution set of problem (2.1) (respectively, problem (2.2)). Note that in the single-valued case, where $G : U \to R^n$, problems (2.1) and (2.2) reduce to problems (1.1) and (1.2), respectively. Recall some definitions of monotonicity type properties for multivalued mappings.

Definition 2.1.2. *[24, 53, 192, 139, 40] Let W and V be convex sets in R^n, $W \subseteq V$, and let $Q : V \to \Pi(R^n)$ be a multivalued mapping.* The mapping Q is said to be

(a) *strongly monotone* on W with constant $\tau > 0$ if for each pair of points $u, v \in W$ and for all $q' \in Q(u)$, $q'' \in Q(v)$, we have

$$\langle q' - q'', u - v \rangle \geq \tau \|u - v\|^2;$$

(b) *strictly monotone* on W if for all distinct $u, v \in W$ and for all $q' \in Q(u)$, $q'' \in Q(v)$, we have

$$\langle q' - q'', u - v \rangle > 0;$$

(c) *monotone* on W if for each pair of points $u, v \in W$ and for all $q' \in Q(u)$, $q'' \in Q(v)$, we have

$$\langle q' - q'', u - v \rangle \geq 0;$$

(d) *pseudomonotone* on W if for each pair of points $u, v \in W$ and for all $q' \in Q(u)$, $q'' \in Q(v)$, we have

$$\langle q'', u - v \rangle \geq 0 \quad \text{implies} \quad \langle q', u - v \rangle \geq 0;$$

(e) *quasimonotone* on W if for each pair of points $u, v \in W$ and for all $q' \in Q(u)$, $q'' \in Q(v)$, we have

$$\langle q'', u - v \rangle > 0 \quad \text{implies} \quad \langle q', u - v \rangle \geq 0;$$

(f) *explicitly quasimonotone* on W if it is quasimonotone on W and for all distinct $u, v \in W$ and for all $q' \in Q(u)$, $q'' \in Q(v)$, the relation

$$\langle q'', u - v \rangle > 0$$

implies

$$\langle q, u - v \rangle > 0 \text{ for some } q \in Q(z), z \in (0.5(u + v), u).$$

From the definitions we obviously have the following implications:

$$(a) \implies (b) \implies (c) \implies (d) \implies (e) \text{ and } (f) \implies (e).$$

The reverse assertions are not true in general. Nevertheless, we can state the following.

Lemma 2.1.1. *Each pseudomonotone mapping on U is explicitly quasimonotone on U.*

Proof. Fix $u, v \in U$. In case (i), assume for contradiction that Q is pseudomonotone on U, but $\langle q', u - v \rangle > 0$ for some $q' \in Q(v)$, and for any $z \in (0.5(u + v), u)$ there is $q \in Q(z)$ such that $\langle q, u - v \rangle \leq 0$, hence $\langle q, v - z \rangle \geq 0$. By pseudomonotonicity, for every $q'' \in Q(v)$ we then have $\langle q'', v - z \rangle \geq 0$, hence,

$$\langle q'', u - v \rangle \leq 0,$$

which is a contradiction. □

Now we give an extension of the Minty Lemma for the multivalued case.

Proposition 2.1.3. *[198, 127, 121]*
   (i) The set U^d is convex and closed.
   (ii) If G is u-hemicontinuous and has nonempty convex and compact values, then $U^d \subseteq U^*$.
   (iii) If G is pseudomonotone, then $U^* \subseteq U^d$.

The existence of solutions to DGVI will play a crucial role for convergence of CR methods for GVI. Note that the existence of a solution to (2.2) implies that (2.1) is also solvable under mild assumptions, whereas the reverse assertion needs generalized monotonicity assumptions.

Proposition 2.1.4. *[121, 40] Suppose that $G : U \to \Pi(R^n)$ is explicitly quasimonotone and that there exist a bounded subset W of U and a point $v \in W$ such that, for every $u \in U \setminus W$ and for every $g \in G(v)$, we have*

$$\langle g, v - u \rangle < 0.$$

Then $U^d \neq \emptyset$.

Note that the assertion above does not hold, generally speaking, for quasimonotone mappings; see Example 1.1.2. We now recall the conditions under which GVI (2.1) has a unique solution.

Proposition 2.1.5. *(e.g. [53, 219])*
(i) If G is strictly monotone, then GVI (2.1) has at most one solution.
(ii) If G is a strongly monotone K-mapping, then GVI (2.1) has a unique solution.

2.1.2 Generalized Variational Inequalities and Related Problems

The relationships between GVI's and other general problems of Nonlinear Analysis, which involve multivalued mappings, are in general the same as in the single-valued case. Namely, in the case of $U = R^n$, GVI (2.1) reduces to the following *multivalued inclusion problem*: Find $u^* \in R^n$ such that

$$0 \in G(u^*).$$

Next, suppose that U is a convex cone in R^n and consider the following *generalized complementarity problem* (GCP for short): Find $u^* \in U$ such that

$$\exists g^* \in G(u^*), \quad g^* \in U', \quad \langle g^*, u^* \rangle = 0. \tag{2.3}$$

This problem is in fact a particular case of GVI (2.1).

Proposition 2.1.6. *Let U be a convex cone. Then problem (2.1) is equivalent to problem (2.3).*

This assertion is proved by the argument similar to that in [85].

Let U be again an arbitrary convex closed set in R^n and let T be a multivalued mapping from U into itself. The *multivalued fixed point problem* associated to the mapping T can be defined as follows: Find $u^* \in U$ such that

$$u^* \in T(u^*). \tag{2.4}$$

The fixed point problem (2.4) can be also converted into a GVI format as stated below in the multivalued version of Proposition 1.1.7.

Proposition 2.1.7. *If the mapping G is defined by $G(u) = u - T(u)$, then problem (2.1) coincides with problem (2.4).*

Let us now consider the optimization problem (1.7), where the function f is not necessarily differentiable. We rewrite this problem here for the convenience of the reader. Namely, given a real-valued function $f : U \to R \bigcup \{+\infty\}$, it is necessary to minimize it over U, which can be written briefly as follows:

$$\min \to \{f(u) \mid u \in U\}. \qquad (2.5)$$

The solution set of this problem we denote by U_f. We can now use all the convexity type concepts from Definition 1.1.3, excepting the pseudomonotonicity which depends on the kind of the subdifferential (directional derivative). One of the most general concepts was proposed by F.H. Clarke and extended by R.T. Rockafellar. Namely, the subdifferential of a function f at a point x is defined as follows:

$$\partial f(x) = \{g \in R^n \mid \langle g, p \rangle \le f^\uparrow(x, p)\},$$

where $f^\uparrow(x, p)$ is the upper Clarke-Rockafellar derivative for f at x in the direction p:

$$f^\uparrow(x, p) = \sup_{\varepsilon > 0} \lim_{\substack{y \to_f x \\ \alpha \to 0}} \inf_{d \in B(p, \varepsilon)} ((f(y + \alpha d) - f(y))/\alpha).$$

Here $y \to_f x$ means that both $y \to x$ and $f(y) \to f(x)$. It is known that

$$f^\uparrow(x, p) = \limsup_{y \to x, \alpha \searrow 0} ((f(y + \alpha p) - f(y))/\alpha)$$

in the case where f is Lipschitz continuous in a neighborhood of x. By definition, $\partial f(x)$ is convex and closed, so that we have

$$f^\uparrow(x, p) = \sup_{g \in \partial f(x)} \langle g, p \rangle. \qquad (2.6)$$

Next, if f is convex, then $\partial f(x)$ coincides with the subdifferential in the sense of Convex Analysis, i.e.,

$$\partial f(x) = \{g \in R^n \mid f(y) - f(x) \ge \langle g, y - x \rangle \quad \forall y \in R^n\}.$$

It follows that

$$f'(x, p) = \lim_{\alpha \to 0} ((f(x + \alpha p) - f(x))/\alpha) = \sup_{g \in \partial f(x)} \langle g, p \rangle$$

and that

$$f^\uparrow(x, p) = f'(x, p). \qquad (2.7)$$

Also, if f is differentiable at x, (2.7) obviously holds and, on account of (2.6), we have

$$f'(x, p) = \langle \nabla f(x), p \rangle \text{ and } \partial f(x) = \{\nabla f(x)\}. \qquad (2.8)$$

Definition 2.1.3. Let W and V be convex sets in R^n, $W \subseteq V$, and let and let $\varphi : V \to R\bigcup\{+\infty\}$ be a function. The function φ is said to be

(a) *proper* if there exists a point $u \in V$ such that $\varphi(u) < \infty$;

(b) *subdifferentiable* at $u \in V$ if $\partial\varphi(u) \neq \emptyset$;

(c) *upper (lower) semicontinuous* (u.s.c. (l.s.c.)) on W, if for each sequence $\{u^k\} \to \bar{u}$, $u^k \in W$ we have $\limsup_{k\to\infty} \varphi(u^k) \leq \varphi(\bar{u})$ ($\liminf_{k\to\infty} \varphi(u^k) \geq \varphi(\bar{u})$);

(d) *continuous* on W if it is both u.s.c. and l.s.c. on W;

(e) *pseudoconvex* on W, if for each pair of points $u, v \in W$, we have

$$\exists g \in \partial\varphi(v), \quad \langle g, u - v \rangle \geq 0 \quad \text{implies} \quad \varphi(u) \geq \varphi(v);$$

(f) *regular* at $u \in V$, if $\varphi^\uparrow(u, p) = \varphi'(u, p)$ for each $p \in R^n$.

From the definitions it follows that each convex function on W is pseudoconvex on W. It is easy to see that Definition 2.1.3 (e) coincides with Definition 1.1.3 (d) when φ is differentiable. Moreover, if φ is l.s.c. and pseudoconvex on W, then it is semistrictly quasiconvex on W; see [174, 40].

Proposition 2.1.8. *Let $\varphi : V \to R\bigcup\{+\infty\}$ be a function.*

(i) [31, Proposition 2.1.2] If φ is Lipschitz continuous in some neighborhood of u, then it is subdifferentiable at u.

(ii) [31, Proposition 2.2.6] If φ is convex and either continuous at u or bounded in some neighborhood of u, then it is subdifferentiable at u.

(iii) [31, Proposition 2.1.5] If the assumptions of (i) or (ii) hold at any point $u \in V$, then $\partial\varphi$ is a K-mapping on V.

We now state the relationships between (generalized) convexity of functions and (generalized) monotonicity of their subdifferentials.

Proposition 2.1.9. *[31, 214, 73, 139, 7, 174, 40] Let W be an open convex subset of W and let $\varphi : V \to R\bigcup\{+\infty\}$ be a function. The function φ is strongly convex with constant $\tau > 0$ (respectively, strictly convex, convex, pseudoconvex, explicitly quasiconvex, quasiconvex) on W if and only if its subdifferential $\partial\varphi$ is strongly monotone with constant $\tau > 0$ (respectively, strictly monotone, monotone, pseudomonotone, explicitly quasimonotone, quasimonotone) on W.*

Nevertheless, the subdifferential in fact possesses additional monotonicity type properties. Recall that the gradient map of a function enjoys the integrability, so that, in particular, its Jacobian must be symmetric. In the multivalued case, certain cyclic monotonicity type conditions hold for the subdifferentials. For instance, we give such a result in the convex case.

Proposition 2.1.10. *[31, 185] Let W be a convex set in R^n and let $\varphi : W \to R\bigcup\{+\infty\}$ be a proper function. The function φ is convex if and only if its* subdifferential $\partial\varphi$ is *cyclically monotone*, i.e., for arbitrary points $x^i \in W$ and $g^i \in \partial\varphi(x^i)$, $i = 1, \ldots, m$ we have

$$\langle g^1, x^2 - x^1 \rangle + \langle g^2, x^3 - x^2 \rangle + \ldots + \langle g^m, x^1 - x^m \rangle \leq 0.$$

One of the earliest rules of the subdifferential calculus is related to the sum of functions. In the convex case, it is known as the Moreau-Rockafellar rule.

Proposition 2.1.11. *[185, 31, 189] Suppose that $\varphi_i : R^n \to R\bigcup\{+\infty\}$, $i = 1, \ldots, m$ are subdifferentiable functions. Set*

$$\varphi(x) = \sum_{i=1}^{m} \varphi_i(x).$$

Then:
  (i)

$$\partial\varphi(x) \subseteq \sum_{i=1}^{m} \partial\varphi_i(x).$$

  (ii) if φ_1 is a proper convex function which is finite at u, and φ_i, $i = 2, \ldots, m$ are convex functions which are continuous at u; we have

$$\partial\varphi(x) = \sum_{i=1}^{m} \partial\varphi_i(x).$$

The following optimality condition is an extension of the Fermat theorem for the nonsmooth case.

Proposition 2.1.12. *[189, 31] Suppose that u^* is a minimizer for a function $\varphi : R^n \to R\bigcup\{+\infty\}$ over R^n. It φ is subdifferentiable at u^*, then $0 \in \partial\varphi(u^*)$.*

For a convex set W, we can define its indicator function $\delta_W : R^n \to R\bigcup\{+\infty\}$ as follows:

$$\delta_W(v) = \begin{cases} 0 & \text{if } v \in W, \\ +\infty & \text{if } v \notin W. \end{cases}$$

It is clear that δ_W is convex and l.s.c., moreover,

$$\partial\delta_W(v) = \begin{cases} N(W, v) & \text{if } v \in W, \\ \emptyset & \text{if } v \notin W. \end{cases}$$

Next, the constrained minimization problem (2.5) is obviously equivalent to the problem of minimizing the function

$$\varphi(u) = f(u) + \delta_W(u)$$

over R^n. By using Proposition 2.1.12 and the definition of pseudoconvexity we now obtain an extension of Theorem 1.1.1.

Theorem 2.1.1. *Suppose that $f : U \to R \bigcup \{+\infty\}$ is a subdifferentiable function. Then:*

(i) $U_f \subseteq U^*$, i.e., each solution to (2.5) is a solution to VI (2.1); where

$$G(u) = \partial f(u); \tag{2.9}$$

(ii) if f is pseudoconvex and G is defined by (2.9), then $U^* \subseteq U_f$.

On account of (2.8), Theorem 2.1.1 reduces to Theorem 1.1.1 when f is differentiable. Thus, optimization problem (2.5) can be reduced to GVI (2.1), with G having additional properties of a cyclic monotonicity type, due to (2.9) and Proposition 2.1.10, for instance.

Theorem 2.1.1 allows us to obtain existence and uniqueness results for optimization problems with nondifferentiable functions.

Corollary 2.1.1. *Let W be an open convex set in R^n such that $U \subseteq W$. Let $f : W \to R \bigcup \{+\infty\}$ be a subdifferentiable, strongly convex function. Then there exists a unique solution to problem (2.5).*

Proof. From Theorem 2.1.1 it follows that problems (2.5) and (2.1), (2.9) are equivalent, moreover, G is strongly monotone due to Proposition 2.1.9. The desired result now follows from Proposition 2.1.5 (ii). □

Let us now consider the optimization problem (1.13) – (1.15) where $f_i : R^l \to R$, $i = 0, \ldots, m$ are convex and subdifferentiable. Then, Proposition 1.1.10 remains valid; e.g., see [185, Chapter 6]. Let $F : R^n \to \Pi(R^n)$ be a multivalued mapping. Then we can define the following GVI: Find $x^* \in D$ such that

$$\exists f^* \in F(x^*), \quad \langle f^*, x - x^* \rangle \geq 0 \quad \forall x \in D, \tag{2.10}$$

where D is defined by (1.14), (1.15).

Proposition 2.1.13. *(i) If $u^* = (x^*, y^*)$ is a solution to GVI (2.1) with*

$$U = X \times Y, Y = R_+^m, G(u) = G(x, y) = \left(\begin{array}{c} F(x) + \sum_{i=1}^{m} y_i \partial f_i(x) \\ -f(x) \end{array} \right), \tag{2.11}$$

then x^* is a solution to problem (2.10), (1.14), (1.15).

(ii) If condition (C) of Section 1.1 holds and x^* is a solution to problem (2.10), (1.14), (1.15), then there exists a point $y^* \in Y = R_+^m$ such that (x^*, y^*) is a solution to (2.1), (2.11).

The proof of these assertions is similar to that of Proposition 1.1.11 with using Theorem 2.1.1 instead of Theorem 1.1.1. Thus, the nonlinear constrained problem (2.10) can be also reduced to GVI (or GCP) with simple constraints, under rather wild assumptions. It should be also noted that G in (2.11) is monotone; see [185, Section 37].

2.1.3 Equilibrium and Mixed Variational Inequality Problems

We now consider other general problems of Nonlinear Analysis and their relationships with several classes of GVI's.

Definition 2.1.4. A bifunction $\Phi : U \times U \to R \bigcup \{+\infty\}$ is called an *equilibrium bifunction*, if $\Phi(u, u) = 0$ for each $u \in U$.

Let U be a nonempty, closed and convex subset of R^n and let $\Phi : U \times U \to R \bigcup \{+\infty\}$ be an equilibrium bifunction. The *equilibrium problem* (EP for short) is the problem of finding a point $u^* \in U$ such that

$$\Phi(u^*, v) \geq 0 \quad \forall v \in U. \tag{2.12}$$

EP gives a suitable and general format for various problems in Game Theory, Mathematical Physics, Economics and other fields. It is easy to see that EP (2.12) can be viewed as some extension of GVI (2.1). In fact, set

$$\Phi(u, v) = \sup_{g \in G(u)} \langle g, v - u \rangle. \tag{2.13}$$

Then each solution of (2.1) is a solution of (2.12), (2.13). The reverse assertion is true if G has convex and compact values; e.g. see [21]. Further, by analogy with the dual variational inequality (2.2), we can define the following *dual equilibrium problem*: Find $v^* \in U$ such that

$$\Phi(u, v^*) \leq 0 \quad \forall u \in U. \tag{2.14}$$

We denote by U^0 and $U^0_{(d)}$ the solution sets of problems (2.12) and (2.14), respectively. The following existence result for EP was obtained by Ky Fan and extended by H. Brézis, L. Nirenberg and G. Stampacchia.

Proposition 2.1.14. *[52, 23] Let $\Phi : U \times U \to R \bigcup \{+\infty\}$ be an equilibrium bifunction such that $\Phi(\cdot, v)$ is u.s.c. for each $v \in U$ and $\Phi(u, \cdot)$ is quasiconvex for each $u \in U$. Suppose also that at least one of the following assumptions hold:*

(a) U is bounded;

(b) there exists a nonempty bounded subset W of U such that for every $u \in U \setminus W$ there is $v \in W$ with $\Phi(u, v) < 0$.

Then EP (2.12) has a solution.

In order to obtain relationships between U^0 and $U^0_{(d)}$, one usually needs monotonicity type conditions on Φ, which can be viewed as some extensions of those in Definition 2.1.2.

Definition 2.1.5. [151, 23, 21, 20, 18] Let W and V be convex sets in R^n, $W \subseteq V$, and let $\Phi : V \times V \to R \bigcup \{+\infty\}$ be an equilibrium bifunction. The bifunction Φ is said to be

(a) *strongly monotone* on W with constant $\tau > 0$ if for each pair of points $u, v \in W$, we have

$$\Phi(u, v) + \Phi(v, u) \leq -\tau \|u - v\|^2;$$

(b) *strictly monotone* on W if for all distinct $u, v \in W$, we have

$$\Phi(u, v) + \Phi(v, u) < 0;$$

(c) *monotone* on W if for each pair of points $u, v \in W$, we have

$$\Phi(u, v) + \Phi(v, u) \leq 0;$$

(d) *pseudomonotone* on W if for each pair of points $u, v \in W$, we have

$$\Phi(u, v) \geq 0 \quad \text{implies } \Phi(v, u) \leq 0;$$

(e) *quasimonotone* on W if for each pair of points $u, v \in W$, we have

$$\Phi(u, v) > 0 \quad \text{implies } \Phi(v, u) \leq 0;$$

(f) *explicitly quasimonotone* on W if it is quasimonotone on W and for all distinct $u, v \in W$, the relation

$$\Phi(u, v) > 0$$

implies

$$\Phi(z, u) < 0 \text{ for some } z \in (0.5(u + v), v).$$

It is easy to see that properties (a) – (f) coincide with the corresponding properties (a) – (f) in Definition 2.1.2 provided (2.13) holds and G has convex and compact values. Besides, from the definitions above we obviously have the following implications:

$$(a) \implies (b) \implies (c) \implies (d) \implies (e) \text{ and } (f) \implies (e).$$

The reverse assertions are not true in general. Nevertheless, under additional continuity assumptions, it is easy to obtain the implication $(d) \implies (f)$.

Lemma 2.1.2. *Let $\Phi : V \times V \to R \bigcup \{+\infty\}$ be an equilibrium bifunction, $\Phi(u, \cdot)$ is l.s.c. for each $u \in V$. If Φ is pseudomonotone on convex set $W \subseteq V$, then it is explicitly quasimonotone on W.*

Proof. Fix $u, v \in W$, $u \neq v$. Assume, for contradiction, that $\Phi(u, v) > 0$ and for any $z \in (0.5(u + v), v)$, we have $\Phi(z, u) \geq 0$. By pseudomonotonicity, it follows that $\Phi(u, z) \leq 0$. Taking the limit $z \to v$ in this inequality gives $\Phi(u, v) \leq 0$, a contradiction. $\qquad \square$

The following assertion is an extension of the Minty Lemma for EP's.

Proposition 2.1.15. *[9, Section 10.1], [18, 116] Let $\Phi : U \times U \to R \bigcup \{+\infty\}$ be an equilibrium bifunction.*

(i) If $\Phi(u, \cdot)$ is quasiconvex and l.s.c. for each $u \in U$, then $U^0_{(d)}$ is convex and closed.

(ii) If $\Phi(\cdot, v)$ is u.s.c. for each $v \in U$, $\Phi(u, \cdot)$ is explicitly quasiconvex for each $u \in U$, then $U^0_{(d)} \subseteq U^0$.

(iii) If Φ is pseudomonotone, then $U^0 \subseteq U^0_{(d)}$.

Also, monotonicity properties enable one to obtain uniqueness results of solutions for EP.

Proposition 2.1.16. *Let $\Phi : U \times U \to R \bigcup \{+\infty\}$ be an equilibrium bifunction.*

(i) If Φ is strictly monotone, then EP (2.12) has at most one solution.

(ii) If $\Phi(\cdot, v)$ is u.s.c. for each $v \in U$, $\Phi(u, \cdot)$ is convex and l.s.c. for each $u \in U$, and Φ is strongly monotone, then EP (2.12) has a unique solution.

Proof. (i) Suppose that u and v are solutions of problem (2.12) and that $u \neq v$. Then $\Phi(u, v) \geq 0$ and $\Phi(v, u) \leq 0$. If Φ is strictly monotone, the first relation implies $\Phi(v, u) < 0$, a contradiction.

(ii) Due to part (i), it suffices to show that EP (2.12) is solvable. Take any $z \in U$. Since $\Phi(z, \cdot)$ is l.s.c., there is a number $\mu > -\infty$ such that

$$\Phi(z, w) \geq \mu \quad \forall w \in B(z, 1) \bigcap U.$$

Choose any $u \in U \setminus B(z, 1)$ and set $\lambda = 1/\|z - u\|$. Then $w = \lambda u + (1 - \lambda)z \in B(z, 1) \bigcap U$ and, by convexity, we have

$$\mu \leq \Phi(z, w) \leq \Phi(z, u)/\|z - u\|.$$

Using the strong monotonicity of Φ now gives

$$\Phi(u, z) \leq -\mu\|z - u\| - \tau\|z - u\|^2 = -\|z - u\|(\mu + \tau\|z - u\|) < 0$$

if $\|z - u\| > -\mu/\tau$. Thus, all the assumptions of Proposition 2.1.14 (b) hold where $W = U \bigcap B(z, \mu')$ and $\mu' > \max\{1, -\mu/\tau\}$. This proves part (ii). $\square$

It is clear that the saddle point problem (1.11) is a particular case of EP. In fact, it suffices to set

$$\Phi(u, v) = L(x', y) - L(x, y'), u = (x, y)^T, v = (x', y')^T \qquad (2.15)$$

and $U = X \times Y$ in (2.12). Note that Φ is obviously monotone and concave-convex in that case. In addition, if $L : X \times Y \to R$ is (strictly) strongly convex in x and (strictly) strongly concave in y, then Φ in (2.15) will be (strictly) strongly concave-convex. Next, the problem of solving a zero-sum two-person game is also a particular case of EP (2.12). Moreover, we can consider the general case of an m-person noncooperative game. Recall that such a game consists of m players, each of which has a strategy set $X_i \subseteq R^{n_i}$ and a utility function $f_i : U \to R$, where

$$U = X_1 \times \ldots \times X_m.$$

A point $u^* = (x_1^*, \ldots, x_m^*)^T \in U$ is said to be a *Nash equilibrium point* for this game, if

$$f_i(x_1^*, \ldots, x_{i-1}^*, y_i, x_{i+1}^*, \ldots, x_m^*)^T) \leq f_i(u^*) \quad \forall y_i \in X_i, i = 1, \ldots, m.$$

Set

$$\Psi(u, v) = -\sum_{i=1}^{m} f_i(x_1, \ldots, x_{i-1}, y_i, x_{i+1}, \ldots, x_m), \qquad (2.16)$$

$$u = (x_1, \ldots, x_m)^T, v = (y_1, \ldots, y_m)^T,$$

and

$$\Phi(u, v) = \Psi(u, v) - \Psi(u, u), \qquad (2.17)$$

then EP (2.12) becomes equivalent to the problem of finding Nash equilibrium points.

Another interesting example is the mixed variational inequality problem. Let $F : U \to R^n$ be a continuous mapping, $\varphi : R^n \to R \bigcup \{+\infty\}$ a l.s.c., proper, convex, but not necessarily differentiable, function. The *mixed variational inequality problem* (MVI for short) is the problem of finding a point $u^* \in U$ such that

$$\langle F(u^*), v - u^* \rangle + \varphi(v) - \varphi(u^*) \geq 0 \quad \forall v \in U. \qquad (2.18)$$

If we set

$$\Phi(u, v) = \langle F(u), v - u \rangle + \varphi(v) - \varphi(u), \qquad (2.19)$$

then problems (2.12) and (2.18) obviously coincide. Therefore, we can obtain existence and uniqueness results for MVI from Propositions 2.1.14 – 2.1.16.

Corollary 2.1.2. *Suppose that at least one of the following assumptions holds:*

(a) U is bounded;

(b) there exists a nonempty bounded subset W of U such that for every $u \in U \backslash W$ there is $v \in W$ with

$$\langle F(u), u - v \rangle + \varphi(u) - \varphi(v) > 0.$$

Then MVI (2.18) has a solution.

Corollary 2.1.3. *(i) If $u^* \in U$ is a solution to the problem*

$$\langle F(u), u - u^* \rangle + \varphi(u) - \varphi(u^*) \geq 0 \quad \forall u \in U, \qquad (2.20)$$

then it is a solution to problem (2.18).

(ii) The solution set of problem (2.20) is convex and closed.

(iii) If F is monotone, then problem (2.18) is equivalent to (2.20).

Proof. On account of (2.19), parts (i) and (ii) follow from parts (ii) and (i) of Proposition 2.1.15, respectively. Next, if F is monotone, then, for all $u, v \in U$, we have

$$
\begin{aligned}
\Phi(u,v) + \Phi(v,u) &= \langle F(u), v - u \rangle + \langle F(v), u - v \rangle \\
&= \langle F(u) - F(v), v - u \rangle \leq 0,
\end{aligned}
$$

i.e., Φ in (2.19) is monotone and part (iii) now follows from part (iii) of Proposition 2.1.15 and part (i).   □

Corollary 2.1.4. *(i) If F is strictly monotone, then MVI (2.18) has at most one solution.*

(ii) If F is strongly monotone, then MVI (2.18) has a unique solution.

Proof. If F is strictly (strongly) monotone, then the equilibrium bifunction Φ in (2.19) is obviously strictly (strongly) monotone. Hence, the desired result now follows from Proposition 2.1.16.   □

From Theorem 2.1.1 it follows that the necessary and sufficient optimality conditions in optimization can be written as GVI. Following this idea, we can make use of GVI to formulate optimality conditions for EP. So, consider EP (2.12) and set

$$
G(u) = \partial \Phi_v(u,v)|_{v=u} . \tag{2.21}
$$

We now establish relationships between monotonicity type properties of Φ and G in (2.21).

Proposition 2.1.17. *Suppose that $\Phi : V \times V \to R \bigcup \{+\infty\}$ is an equilibrium bifunction such that $\Phi(u, \cdot)$ is convex and subdifferentiable for each $u \in U$. If Φ is strongly monotone with constant τ (respectively, strictly monotone, monotone, pseudomonotone, quasimonotone, explicitly quasimonotone) on U, then so is G in (2.21).*

Proof. Take any $u, v \in U$ and $g' \in G(u)$, $g'' \in G(v)$. Then, by convexity,

$$
\Phi(u,v) = \Phi(u,v) - \Phi(u,u) \geq \langle g', v - u \rangle, \tag{2.22}
$$

$$
\Phi(v,u) = \Phi(v,u) - \Phi(v,v) \geq \langle g'', u - v \rangle. \tag{2.23}
$$

Adding these inequalities gives

$$
\Phi(u,v) + \Phi(v,u) \geq \langle g' - g'', v - u \rangle.
$$

Hence, if Φ is strongly monotone with constant τ (respectively, strictly monotone, monotone), so is G. If Φ is pseudomonotone and $\langle g', v - u \rangle \geq 0$ then $\Phi(u,v) \geq 0$ due to (2.22), hence $\Phi(v,u) \leq 0$. Applying this inequality in (2.23) gives $\langle g'', u - v \rangle \leq 0$, i.e., G is pseudomonotone. Analogously, we see that G is quasimonotone, whenever so is Φ. Next, let Φ be explicitly quasimonotone and $\langle g', v - u \rangle > 0$ with $u \neq v$. Then (2.22) gives $\Phi(u,v) > 0$, hence there is $z \in (0.5(u + v), v)$ such that $\Phi(z, u) < 0$. Since

$$\Phi(z, u) \geq \langle g, u - z \rangle \quad \forall g \in G(z),$$

it follows that $\langle g, u - v \rangle < 0$, hence G is also explicitly quasimonotone.   □

Note that the reverse assertions are not true in general (see Example 3.2.7). Applying this result to the saddle point problem (1.11) and using (2.15), we now see that the mapping G in (2.21) will be monotone if L is convex-concave. This is the case if we consider the convex optimization problem (2.5), (1.14), (1.15). The following result gives an additional property of a pseudoconvex function.

Lemma 2.1.3. *[39, Proposition 2.3] Let $f : U \to R \bigcup \{+\infty\}$ be l.s.c. and let ∂f be pseudomonotone. Then, for all $x, y \in U$, we have*

$$\exists g \in \partial f(x), \quad \langle g, y - x \rangle > 0 \quad \Longrightarrow \quad f(y) > f(x).$$

We are now ready to establish an optimality condition for EP.

Theorem 2.1.2. *Let W be an open convex set in R^n such that $U \subseteq W$. Let $\Phi : W \times W \to R \bigcup \{+\infty\}$ be an equilibrium bifunction and let the mapping G be defined by (2.21).*

(i) If $\Phi(u, \cdot)$ is quasiconvex and regular for each $u \in W$, then

$$U^0 \subseteq U^*, U^0_{(d)} \subseteq U^d.$$

(ii) If $\Phi(u, \cdot)$ is pseudoconvex and l.s.c. for each $u \in W$, then

$$U^0 = U^*, U^0_{(d)} \subseteq U^d.$$

Proof. Take any $u^* \in U^0$. Then it is a solution to the following optimization problem

$$\min \to \{\Phi(u^*, v) \mid v \in U\}.$$

From Theorem 2.1.1 (i) it now follows that u^* is a solution to GVI (2.1), (2.21), i.e., $U^0 \subseteq U^*$. Moreover, if $\Phi(u, \cdot)$ is pseudoconvex, then the reverse inclusion holds due to Theorem 2.1.1 (ii). We will now prove the inclusion $U^0_{(d)} \subseteq U^d$ for both cases. Take any $u^* \in U^0_{(d)}$. Then $\Phi(u, u^*) \leq \Phi(u, u) = 0$ for each $u \in U$. Set $\psi(v) = \Phi(u, v)$. In case (i), we have

$$\psi^\uparrow(u, u^* - u) = \psi'(u, u^* - u) = \lim_{\alpha \to 0}[(\psi(u + \alpha(u^* - u)) - \psi(u))/\alpha] \leq 0,$$

hence, by definition,

$$\langle g, u^* - u \rangle \leq \psi^\uparrow(u, u^* - u) \leq 0 \quad \forall g \in \partial \psi(u).$$

But $\partial \psi(u) = G(u)$, therefore $u^* \in U^d$. Next, in case (ii), ψ is l.s.c. and pseudoconvex, hence $\partial \psi$ is pseudomonotone due to Proposition 2.1.9. Since $\psi(u^*) \leq \psi(u)$, we must have

$$\langle g, u^* - u \rangle \leq 0 \quad \forall g \in \partial \psi(u)$$

because of Lemma 2.1.3, hence $u^* \in U^d$.   □

Theorem 2.1.2 allows us to obtain additional properties of MVI (2.18), including existence and uniqueness results.

Proposition 2.1.18. *Let $F : U \to R^n$ be a continuous mapping, $\varphi : R^n \to R \bigcup \{+\infty\}$ a l.s.c., proper, convex and subdifferentiable function.*
(i) Problem (2.18) is equivalent to the problem of finding $u^* \in U$ such that

$$\exists t^* \in \partial\varphi(u^*), \quad \langle F(u^*) + t^*, u - u^* \rangle \geq 0 \quad \forall u \in U. \tag{2.24}$$

(ii) If φ is strictly convex, then MVI (2.18) has at most one solution.
(iii) If φ is strongly convex, then MVI (2.18) has a unique solution.

Proof. Since MVI (2.18) is equivalent to EP (2.12) with Φ being defined in (2.19), we see that all the assumptions of Theorem 2.1.2 (ii) hold, moreover,

$$G(u^*) = F(u^*) + \partial\varphi(u^*) \tag{2.25}$$

due to Proposition 2.1.11. Therefore, (2.18) is equivalent to (2.24). Next, if φ is strictly (strongly) convex, then $\partial\varphi$ is strictly (strongly) monotone due to Proposition 2.1.9, hence so is G in (2.25). The assertions of parts (ii) and (iii) now follow from Proposition 2.1.5 and part (i).   □

2.2 CR Method for the Mixed Variational Inequality Problem

In this section, we describe a CR method, which follows Basic Scheme I, for MVI (2.18). Its auxiliary procedure is based on splitting techniques [138, 171]. Throughout this section, we suppose that $F : V \to R$ is a locally Lipschitz continuous and monotone mapping, $\varphi : R^n \to R$ is a convex function, V is a convex closed set in R^n such that $U \subseteq V$. Besides, we suppose that MVI (2.18) is solvable.

2.2.1 Description of the Method

Let us first define an auxiliary function $\psi : V \to R$ satisfying the following assumptions.

(B1) *ψ is differentiable and its gradient $\nabla \psi$ is Lipschitz continuous with constant $\tau'' < \infty$;*

(B2) *ψ is strongly convex with constant $\tau' > 0$.*

It is clear that the class of functions satisfying (B1) and (B2) is rather broad; it contains, in particular, the strongly convex quadratic functions. We intend to make use of such functions for constructing a CR method for solving MVI (2.18).

Method 2.1. *Step 0 (Initialization)*: Choose a point $u^0 \in V$, a sequence $\{\gamma_k\}$ such that

$$\gamma_k \in [0, 2], k = 0, 1, \dots; \quad \sum_{k=0}^{\infty} \gamma_k (2 - \gamma_k) = \infty; \qquad (2.26)$$

and a function $\psi : V \to R$ satisfying (B1) and (B2). Choose a sequence of mappings $\{P_k\}$, where $P_k \in \mathcal{F}(V_k)$ and $U \subseteq V_k \subseteq V$ for $k = 0, 1, \dots$. Choose numbers $\alpha \in (0, 1)$, $\beta \in (0, 1)$, $\tilde{\theta} > 0$. Set $k := 0$.

Step 1 (Auxiliary procedure D_k):

Step 1.1 : Find m as the smallest number in Z_+ such that

$$\langle F(u^k) - F(z^{k,m}), u^k - z^{k,m} \rangle \leq (1 - \alpha)(\tilde{\theta}\beta^m)^{-1}$$
$$\langle \nabla\psi(z^{k,m}) - \nabla\psi(u^k), z^{k,m} - u^k \rangle, \qquad (2.27)$$

where $z^{k,m}$ is a solution of the following auxiliary optimization problem:

$$\min \to \{\langle F(u^k) - (\tilde{\theta}\beta^m)^{-1}\nabla\psi(u^k), v \rangle$$
$$+ (\tilde{\theta}\beta^m)^{-1}\psi(v) + \varphi(v) \mid v \in U\}. \qquad (2.28)$$

Step 1.2: Set $\theta_k := \beta^m \tilde{\theta}$, $v^k := z^{k,m}$. If $u^k = v^k$, stop.
Step 2 (Main iteration): Set

$$g^k := F(v^k) - F(u^k) - \theta_k^{-1}(\nabla\psi(v^k) - \nabla\psi(u^k)), \qquad (2.29)$$

$$\sigma_k := \langle g^k, u^k - v^k \rangle / \|g^k\|^2, \qquad (2.30)$$

$$u^{k+1} := P_k(u^k - \gamma_k\sigma_k g^k), \qquad (2.31)$$

$k := k + 1$ and go to Step 1.

Note that the sets V_k and V can be determined rather arbitrarily within $U \subseteq V_k \subseteq V \subseteq R^n$ and that $\{u^k\}$ need not be feasible if $U \subset V$. Indeed, Method 2.1 can be viewed as a modification and an extension of Method 1.4 to the case of MVI (2.18), so that it will possess the same properties. On the other hand, the solution of problem (2.28) can be regarded as a modification of an iteration of the splitting method [138] and [173]. Let us now consider properties of problem (2.28) in more detail.

Lemma 2.2.1. *(i) Problem (2.28) is equivalent to each of the following problems:*

(a) Find $\bar{z} \in U$ such that

$$\langle F(u^k), v - \bar{z} \rangle + (\tilde{\theta}\beta^m)^{-1}\left[\psi(v) - \psi(\bar{z}) - \langle \nabla\psi(u^k), v - \bar{z} \rangle\right]$$
$$+ \varphi(v) - \varphi(\bar{z}) \geq 0 \quad \forall v \in U; \qquad (2.32)$$

(b) find $\bar{z} \in U$ such that

$$\langle F(u^k), v - \bar{z} \rangle + (\tilde{\theta}\beta^m)^{-1}\langle \nabla\psi(\bar{z}) - \nabla\psi(u^k), v - \bar{z} \rangle$$
$$+ \varphi(v) - \varphi(\bar{z}) \geq 0 \quad \forall v \in U; \qquad (2.33)$$

(c) find $\bar{z} \in U$ such that

$$\exists \bar{g} \in \partial\varphi(\bar{z}) : \langle F(u^k) + \bar{g}, v - \bar{z} \rangle$$
$$+ (\theta\beta^m)^{-1}\langle \nabla\psi(\bar{z}) - \nabla\psi(u^k), v - \bar{z} \rangle \geq 0 \quad \forall v \in U. \qquad (2.34)$$

(ii) Problem (2.28) has a unique solution.

Proof. First we observe that (2.28) is equivalent to GVI (2.34) due to Theorem 2.1.1 since the cost function in (2.28) is clearly convex. In turn, (2.34) is equivalent to (2.33) due to Proposition 2.1.18 with using $F(u^k) + (\theta\beta^m)^{-1}(\nabla\psi(z) - \nabla\psi(u^k))$ instead of $F(z)$. Set

$$\Phi(u, v) = \langle F(u^k), v - u \rangle + (\theta\beta^m)^{-1}\left[\psi(v) - \psi(u) - \langle \nabla\psi(u^k), v - u \rangle\right]$$
$$+ \varphi(v) - \varphi(u).$$

It is clear that Φ is an equilibrium function, $\Phi(u, \cdot)$ is convex and subdifferentiable, so that

$$\partial\Phi_v(u, v)|_{v=u} = F(u^k) + \partial\varphi(u) + (\theta\beta^m)^{-1}(\nabla\psi(u) - \nabla\psi(u^k)).$$

Applying Theorem 2.1.2 (ii), we now conclude that (2.34) is equivalent to (2.32). This proves part (i). Part (ii) now follows from Corollary 2.1.1 since the cost function in (2.28) is strongly convex.   □

Thus, we can replace the auxiliary optimization problem (2.28) in Method 2.1 with any of problems (2.32) – (2.34), which will be most suitable for numerical implementation. Nevertheless, the convex optimization problem (2.28) seems simpler than GVI (2.34) or MVI's (2.32) and (2.33). Next, if we set

$$\psi(u) = 0.5 \, \|u\|^2 \, ,$$

then the solution of problem (2.33) is clearly equivalent to an iteration of the simplest splitting method for MVI (2.18); see [138]. Besides, various fixed point and descent type methods involving a solution of auxiliary problems similar to (2.32), were considered in [163, 172, 173]. However, such methods are convergent if, additionally, either F is strictly (strongly) monotone, or φ is strictly convex. In this section we intend to show that our Method 2.1 need not additional assumptions for convergence.

2.2.2 Convergence

We first show that the linesearch procedure in Step 1 is well defined.

Lemma 2.2.2. *It holds that*
   (i) $\theta_k > 0$, i.e., the linesearch procedure in Step 1 is always finite;
and
   (ii) if $\{u^k\}$ is bounded, then so is $\{v^k\}$ and

$$\theta_k \geq \theta' > 0, \quad k = 0, 1, \ldots \tag{2.35}$$

Proof. Take any $\tilde{u} \in U$. Taking into account Lemma 2.2.1 (i) and applying (2.33) with $v = \tilde{u}$, $\bar{z} = z^{k,m}$ gives

$$\langle F(u^k) + (\tilde{\theta}\beta^m)^{-1}[\nabla\psi(z^{k,m}) - \nabla\psi(u^k)], \tilde{u} - z^{k,m}\rangle + \varphi(\tilde{u}) - \varphi(z^{k,m}) \geq 0.$$

Take any subgradient $\tilde{g}$ for φ at $\tilde{u}$. Then, by the properties of $\tilde{g}$ and ψ, we have

$$
\begin{aligned}
\tilde{\theta}(\|F(u^k)\| \; + \; & \|\tilde{g}\|)\|\tilde{u} - z^{k,m}\| \geq (\tilde{\theta}\beta^m)(\langle F(u^k), \tilde{u} - z^{k,m}\rangle + \langle \tilde{g}, \tilde{u} - z^{k,m}\rangle) \\
\geq \; & (\tilde{\theta}\beta^m)(\langle F(u^k), \tilde{u} - z^{k,m}\rangle + \varphi(\tilde{u}) - \varphi(z^{k,m})) \\
\geq \; & \langle \nabla\psi(z^{k,m}) - \nabla\psi(\tilde{u}), z^{k,m} - \tilde{u}\rangle \\
+ \; & \langle \nabla\psi(\tilde{u}) - \nabla\psi(u^k), z^{k,m} - \tilde{u}\rangle \\
\geq \; & \tau'\|z^{k,m} - \tilde{u}\|^2 - \tau''\|u^k - \tilde{u}\|\|z^{k,m} - \tilde{u}\|.
\end{aligned}
$$

It follows that

$$\|u^k - z^{k,m}\| \leq \|u^k - \tilde{u}\| + \|\tilde{u} - z^{k,m}\|$$
$$\leq \|u^k - \tilde{u}\|(1 + \tau''/\tau') + \tilde{\theta}(\|F(u^k)\| + \|\tilde{g}\|)/\tau' = \lambda_k.$$

Let L_k be the Lipschitz constant for F on the set $\{u \in V \mid \|u - u^k\| \leq \lambda_k\}$. Then, using the properties of ψ gives

$$\langle F(u^k) - F(z^{k,m}), u^k - z^{k,m} \rangle \leq L_k \|u^k - z^{k,m}\|^2$$
$$\leq L_k \langle \nabla\psi(z^{k,m}) - \nabla\psi(u^k), z^{k,m} - u^k \rangle/\tau'$$
$$\leq (1 - \alpha)(\tilde{\theta}\beta^m)^{-1} \langle \nabla\psi(z^{k,m}) - \nabla\psi(u^k), z^{k,m} - u^k \rangle,$$

when $(1 - \alpha)(\tilde{\theta}\beta^m)^{-1} \geq L_k/\tau'$. Taking into account (2.27) we see that

$$\theta_k \geq \min\{\beta(1 - \alpha)\tau'/L_k, \tilde{\theta}\} > 0, \tag{2.36}$$

hence assertion (i) is true. In case (ii), the sequence $\{F(u^k)\}$ is also bounded, hence $\lambda_k \leq \lambda' < \infty$ and $L_k \leq L' < \infty$ for $k = 0, 1, \ldots$ Therefore, (2.36) now implies (2.35) where $\theta' = \min\{\beta(1 - \alpha)\tau'/L', \tilde{\theta}\}$ and the proof is complete. $\square$

Lemma 2.2.3. *(i) If $u^k = v^k$, then u^k solves MVI (2.18).*
(ii) If $u^k \neq v^k$, then

$$\langle g^k, u^k - v^k \rangle \geq (\alpha/\theta_k)\langle \nabla\psi(v^k) - \nabla\psi(u^k), v^k - u^k \rangle$$
$$\geq (\alpha\tau'/\theta_k)\|u^k - v^k\|^2. \tag{2.37}$$

Proof. We first note that $\theta_k > 0$ due to Lemma 2.2.2 (i). Therefore, if $u^k = v^k$, then Lemma 2.2.1 (i) and (2.33) yield

$$\langle F(u^k), v - u^k \rangle + \varphi(v) - \varphi(u^k)$$
$$= \langle F(u^k) + \theta_k^{-1}[\nabla\psi(u^k) - \nabla\psi(u^k)], v - u^k \rangle$$
$$+ \varphi(v) - \varphi(u^k) \geq 0$$

for every $v \in U$, i.e., u^k solves MVI (2.18) and part (i) is true. Next, in case $u^k \neq v^k$, by (2.27), (2.29) and Proposition 1.1.9, we have

$$\langle g^k, u^k - v^k \rangle = \langle F(v^k) - F(u^k) - \theta_k^{-1}[\nabla\psi(v^k) - \nabla\psi(u^k)], u^k - v^k \rangle$$
$$\geq ((1 - \alpha)/\theta_k)\langle \nabla\psi(v^k) - \nabla\psi(u^k), u^k - v^k \rangle$$
$$- (1/\theta_k)\langle \nabla\psi(v^k) - \nabla\psi(u^k), u^k - v^k \rangle$$
$$= (\alpha/\theta_k)\langle \nabla\psi(v^k) - \nabla\psi(u^k), v^k - u^k \rangle$$
$$\geq (\alpha\tau'/\theta_k)\|u^k - v^k\|^2,$$

i.e., part (ii) is also true. $\square$

Note that $u^k \neq v^k$ implies $g^k \neq 0$ due to (2.37), so that Step 2 is well defined.

Lemma 2.2.4. *(i) If u^* solves MVI (2.18), then*

$$\|u^{k+1} - u^*\|^2 \leq \|u^k - u^*\|^2 - \gamma_k(2 - \gamma_k)(\sigma_k\|g^k\|)^2.$$

(ii) $\{u^k\}$ is bounded.

(iii) $\sum\limits_{k=0}^{\infty} \gamma_k(2 - \gamma_k)(\sigma_k\|g^k\|)^2 < \infty$.

(iv) For each limit point u^* of $\{u^k\}$ such that u^* solves MVI (2.18), we have

$$\lim_{k \to \infty} u^k = u^*.$$

Proof. Let $u^* \in U^*$ be a solution to MVI (2.18). Then, by the monotonicity of F, we have

$$\langle F(v^k), v^k - u^* \rangle + \varphi(v^k) - \varphi(u^*) \geq \langle F(u^*), v^k - u^* \rangle + \varphi(v^k) - \varphi(u^*) \geq 0.$$

On account of Lemma 2.2.1 (i), applying (2.33) with $v = u^*$, $\bar{z} = z^{k,m} = v^k$ yields

$$\langle -F(u^k) - \theta_k^{-1}[\nabla\psi(v^k) - \nabla\psi(u^k)], v^k - u^* \rangle + \varphi(u^*) - \varphi(v^k) \geq 0.$$

Adding both the inequalities above gives $\langle g^k, v^k - u^* \rangle \geq 0$ whence

$$\langle g^k, u^k - u^* \rangle = \langle g^k, u^k - v^k \rangle + \langle g^k, v^k - u^* \rangle \geq \sigma_k\|g^k\|^2 \geq 0$$

due to (2.37). As in the proof of Lemma 1.2.3, this relation proves part (i) due to (2.26), (2.30) and (2.31). Parts (ii) – (iv) follow directly from (i). □

We are now in a position to establish a convergence result for Method 2.1.

Theorem 2.2.1. *Let a sequence $\{u^k\}$ be constructed by Method 2.1. If the method terminates at the kth iteration, then u^k solves MVI (2.18). Otherwise, if $\{u^k\}$ is infinite, then $\{u^k\}$ converges to a solution of MVI (2.18).*

Proof. The first assertion follows from Lemma 2.2.3 (i). Let $\{u^k\}$ be infinite. Then, by Lemmas 2.2.4 (ii) and 2.2.2 (ii), $\{u^k\}$, $\{F(u^k)\}$, $\{v^k\}$, and $\{F(v^k)\}$ are bounded. Next, by (2.29), (2.30) and (2.37), we have

$$\begin{aligned}
\sigma_k\|g^k\| &= \langle g^k, u^k - v^k \rangle / \|F(v^k) - F(u^k) - \theta_k^{-1}[\nabla\psi(v^k) - \nabla\psi(u^k)]\| \\
&\geq \alpha\tau'\|u^k - v^k\|^2 / (\theta_k(L + \tau''/\theta_k)\|u^k - v^k\|) \\
&\geq \alpha\tau'\|u^k - v^k\| / (\tilde{\theta}L + \tau''),
\end{aligned} \tag{2.38}$$

where L is the corresponding Lipschitz constant for F. On the other hand, (2.26) and Lemma 2.2.4 (iii) imply that

$$\liminf_{k \to \infty}(\sigma_k\|g^k\|) = 0.$$

It now follows that

$$\liminf_{k\to\infty} \|u^k - v^k\| = 0. \tag{2.39}$$

Since $\{u^k\}$ and $\{v^k\}$ are bounded, they have limit points. Without loss of generality we can conclude that there exist subsequences $\{u^{k_s}\}$ and $\{v^{k_s}\}$ such that

$$\lim_{s\to\infty} u^{k_s} = \lim_{s\to\infty} v^{k_s} = u^* \in U.$$

Take any $v \in U$. Taking into account Lemma 2.2.1 (i), from (2.33) with $\bar{z} = v^{k_s}$, (2.35), and properties of ψ, we have

$$\begin{aligned}
\langle F(u^{k_s}), v - v^{k_s}\rangle + \varphi(v) - \varphi(v^{k_s}) &\geq -\|\nabla\psi(v^{k_s}) - \nabla\psi(u^{k_s})\|\|v - v^{k_s}\|/\theta_{k_s} \\
&\geq -\tau''\|u^{k_s} - v^{k_s}\|\|v - v^{k_s}\|/\theta'.
\end{aligned}$$

Taking the limit $s \to \infty$ in this inequality gives

$$\langle F(u^*), v - u^*\rangle + \varphi(v) - \varphi(u^*) \geq 0,$$

i.e., u^* solves MVI (2.18). The desired result now follows from Lemma 2.2.4 (iv). □

2.2.3 Rate of Convergence

On account of Lemma 2.2.3 and properties of ψ, the value of $\langle\nabla\psi(v^k) - \nabla\psi(u^k), v^k - u^k\rangle$ can be regarded as an error bound for Method 2.1. The following result is based on this observation.

Theorem 2.2.2. *Let an infinite sequence $\{u^k\}$ be generated by Method 2.1. Suppose that*

$$\gamma_k = \gamma \in (0,2), \quad k = 0,1,\dots \tag{2.40}$$

Then

$$\liminf_{k\to\infty} \left(\langle\nabla\psi(v^k) - \nabla\psi(u^k), v^k - u^k\rangle\sqrt{k+1}\right) = 0.$$

Proof. From Lemmas 2.2.4 (ii) and 2.2.2 (ii) we have that sequences $\{u^k\}$, $\{v^k\}$ and $\{G(v^k)\}$ are bounded. Combining (2.37), (2.38) and (B2) now gives

$$\begin{aligned}
(\sigma_k\|g^k\|)^2 &\geq \alpha^2\langle\nabla\psi(v^k) - \nabla\psi(u^k), v^k - u^k\rangle^2/\left((\tilde{\theta}L + \tau'')\|u^k - v^k\|\right)^2 \\
&\geq \alpha^2\langle\nabla\psi(v^k) - \nabla\psi(u^k), v^k - u^k\rangle/\left(\tilde{\theta}L + \tau''\right)^2.
\end{aligned}$$

The rest is proved by the argument similar to that in Theorem 1.4.2 (i). □

According to Corollary 2.1.4 (ii), MVI (2.18) has a unique solution if F is strongly monotone. We will show that our method converges at least linearly under this assumption.

Theorem 2.2.3. *Suppose that F is strongly monotone with constant κ and that (2.40) holds. If Method 2.1 generates an infinite sequence $\{u^k\}$, then $\{u^k\}$ converges to a solution of problem (2.18) in a linear rate.*

Proof. We first show that there exists a number $\mu > 0$ such that

$$\|u^k - v^k\|^2 \geq \mu \|u^k - u^*\|^2, \qquad (2.41)$$

where u^* is a unique solution of problem (2.18). Indeed, from (2.18) with $v = v^k$ we have

$$\langle F(u^*), v^k - u^* \rangle + \varphi(v^k) - \varphi(u^*) \geq 0.$$

On account of Lemma 2.2.1 (i), applying (2.33) with $\bar{z} = v^k$ and $v = u^*$ gives

$$\langle F(u^k) + \theta_k^{-1}[\nabla\psi(v^k) - \nabla\psi(u^k)], u^* - v^k \rangle + \varphi(u^*) - \varphi(v^k) \geq 0.$$

Adding both inequalities yields

$$\langle F(u^*) - F(u^k), v^k - u^k \rangle + \langle F(u^*) - F(u^k), u^k - u^* \rangle$$
$$+\theta_k^{-1} \langle \nabla\psi(v^k) - \nabla\psi(u^k), u^* - v^k \rangle \geq 0,$$

i.e.,

$$\langle F(u^k) - F(u^*), u^k - u^* \rangle \leq \langle F(u^*) - F(u^k), v^k - u^k \rangle$$
$$+\theta_k^{-1} \langle \nabla\psi(v^k) - \nabla\psi(u^k), u^* - v^k \rangle.$$

Taking into account Lemmas 2.2.4 (ii) and 2.2.2 (ii) and the properties of ψ, we have

$$
\begin{aligned}
\kappa\|u^k - u^*\|^2 &\leq \langle F(u^k) - F(u^*), u^k - u^* \rangle \\
&\leq \langle F(u^*) - F(u^k), v^k - u^k \rangle \\
&\quad +\theta_k^{-1} \langle \nabla\psi(v^k) - \nabla\psi(u^k), u^* - v^k \rangle \\
&\leq (L + \tau''/\theta_k)\|u^* - u^k\|\|v^k - u^k\| \\
&\leq (L + \tau''/\theta')\|u^* - u^k\|\|v^k - u^k\|,
\end{aligned}
$$

where L is the corresponding Lipschitz constant for F, i.e., (2.41) holds with $\mu = (\kappa/(L + \tau''/\theta'))^2$. Using (2.38), we then obtain

$$
\begin{aligned}
(\sigma_k\|g^k\|)^2 &\geq (\alpha\tau'\|u^k - v^k\|/(\tilde{\theta}L + \tau''))^2 \\
&\geq (\alpha\tau')^2\|u^k - u^*\|^2\mu/(\tilde{\theta}L + \tau'')^2 = \mu'\|u^k - u^*\|^2.
\end{aligned}
$$

This inequality together with (2.40) and Lemma 2.2.4 (i) gives

$$\|u^{k+1} - u^*\|^2 \leq \|u^k - u^*\|^2 - \gamma(2-\gamma)\mu'\|u^k - u^*\|^2 = (1 - \gamma(2-\gamma)\mu')\|u^k - u^*\|^2,$$

and the result follows. $\qquad\square$

We now give conditions that ensure finite termination of Method 2.1. Let us consider the following assumption, which can be viewed as an extension of (A6).

(B3) *There exists a number $\mu' > 0$ such that, for each point $u \in U$, the following inequality holds:*

$$\langle F(u), u - \pi_{U^*}(u) \rangle + \varphi(u) - \varphi(\pi_{U^*}(u)) \geq \mu' \|u - \pi_{U^*}(u)\|.$$

Theorem 2.2.4. *Let a sequence $\{u^k\}$ be constructed by Method 2.1. Suppose that (B3) holds. Then the method terminates with a solution.*

Proof. Assume for contradiction that $\{u^k\}$ is infinite. Then, following the proof of Theorem 2.2.1, we see that (2.39) holds, moreover, sequences $\{u^k\}$ and $\{v^k\}$ are bounded. On the other hand, from (B1) – (B3), Lemmas 2.2.1 and 2.2.2 (ii) it follows that

$$
\begin{aligned}
\mu' \|v^k - \pi_{U^*}(v^k)\| &\leq \langle F(v^k), v^k - \pi_{U^*}(v^k) \rangle + \varphi(v^k) - \varphi(\pi_{U^*}(v^k)) \\
&= \langle F(v^k) - F(u^k) \\
&\quad -\theta_k^{-1}(\nabla\psi(v^k) - \nabla\psi(u^k)), v^k - \pi_{U^*}(v^k) \rangle \\
&\quad + \langle F(u^k) + \theta_k^{-1}(\nabla\psi(v^k) - \nabla\psi(u^k)), v^k - \pi_{U^*}(v^k) \rangle \\
&\quad + \varphi(v^k) - \varphi(\pi_{U^*}(v^k)) \\
&\leq \langle F(v^k) - F(u^k) \\
&\quad -\theta_k^{-1}(\nabla\psi(v^k) - \nabla\psi(u^k)), v^k - \pi_{U^*}(v^k) \rangle \\
&\leq (L\|v^k - u^k\| + \theta_k^{-1}\tau''\|v^k - u^k\|) \|v^k - \pi_{U^*}(v^k)\| \\
&\leq (L + \tau''/\theta')\|v^k - u^k\|\|v^k - \pi_{U^*}(v^k)\|,
\end{aligned}
$$

where L is the corresponding Lipschitz constant for F. It follows that

$$\|v^k - u^k\| \geq \mu'/(L + \tau''/\theta'),$$

a contradiction.

$\square$

Thus, our method for MVI in general maintains properties of solution methods for VI with continuous mappings.

2.3 CR Method for the Generalized Variational Inequality Problem

We now consider a method for solving GVI (2.1). Since we has to take into account the situation of a null step in the multivalued case, the method in general follows Basic Scheme II from Section 1.6 with corresponding modifications. The blanket assumptions of this section are the following:

Hypothesis (H2.3)

(a) U is a subset of R^n, which is defined by

$$U = \{u \in R^n \mid h(u) \le 0\}, \qquad (2.42)$$

where $h : R^n \to R$ is a convex, but not necessarily differentiable, function;
(b)   the Slater condition is satisfied, i.e., there exists a point $\bar{u}$ such that $h(\bar{u}) < 0$;
(c) $G : U \to \Pi(R^n)$ is a K-mapping;
(d) $U^d \ne \emptyset$.

Note that the feasible set is usually defined by

$$U = \{u \in R^n \mid h_i(u) \le 0 \quad i = 1, \ldots, m\}$$

where $h_i : R^n \to R, i = 1, \ldots, m$ are convex functions. But, by setting $h(u) = \max_{i=1,\ldots,m} h_i(u)$ we can easily reduce this case to the initial (2.42) without loss of generality.

2.3.1 Description of the Method

Let us define the mapping $Q : R^n \to \Pi(R^n)$ by

$$Q(u) = \begin{cases} G(u) & \text{if } h(u) \le 0, \\ \partial h(u) & \text{if } h(u) > 0. \end{cases} \qquad (2.43)$$

Method 2.2. *Step 0 (Initialization):* Choose a point $u^0 \in U$, a sequence $\{\gamma_j\}$ such that

$$\gamma_j \in [0, 2], j = 0, 1, \ldots; \sum_{j=0}^{\infty} \gamma_j (2 - \gamma_j) = \infty; \qquad (2.44)$$

and bounded positive sequences $\{\varepsilon_l\}$ and $\{\eta_l\}$. Also, choose a number $\theta \in (0, 1)$, and a sequence of mappings $\{P_k\}$, where $P_k \in \mathcal{F}(U)$ for $k = 0, 1, \ldots$ Set $k := 0, l := 1, y^0 := u^0, j(0) := 0, k(0) := 0, t := 0, K_l := \{0\}$.
    Step 1 (Auxiliary procedure $\tilde{D}_k$) :
        Step 1.1 : Choose q^0 from $Q(u^k)$, set $i := 0, p^i := q^i, I_{k,l} := \{0\}$, $t := t + 1, w^0 := u^k$.

Step 1.2: If

$$\|p^i\| \leq \eta_l, \tag{2.45}$$

set $k(l) := k$, $y^l := u^{k+1} := u^k$, $j(k+1) := j(k)$, $k := k+1$, $l := l+1$, $K_l := \{k\}$ and go to Step 1. (*null step*)

Step 1.3: Set $w^{i+1} := u^k - \varepsilon_l p^i / \|p^i\|$, choose $q^{i+1} \in Q(w^{i+1})$, set $I_{k,l} := I_{k,l} \bigcup \{i+1\}$, $t := t+1$. If

$$\langle q^{i+1}, p^i \rangle > \theta \|p^i\|^2, \tag{2.46}$$

then set $v^k := w^{i+1}$, $g^k := q^{i+1}$, $j(k+1) := j(k) + 1$, and go to Step 2. (*descent step*)

Step 1.4: Set

$$p^{i+1} := \text{Nr conv}\{p^i, q^{i+1}\}, \tag{2.47}$$

$i := i+1$ and go to Step 1.2.

Step 2 (Main iteration): Set $\sigma_k := \langle g^k, u^k - v^k \rangle / \|g^k\|^2$,

$$u^{k+1} := P_k(u^k - \gamma_{j(k)} \sigma_k g^k),$$

$k := k+1$, $K_l := K_l \bigcup \{k\}$ and go to Step 1.

According to the description, at each iteration, the auxiliary procedure in Step 1 is applied for direction finding. In the case of a null step, the tolerances ε_l and η_l decrease since the point u^k approximates a solution within ε_l, η_l (see Lemma 2.3.4 (ii) below). Hence, the variable l is a counter for null steps and the variable $j(\cdot)$ is a counter for descent steps. In the case of a descent step we must have $\sigma_k > 0$ (see Lemma 2.3.4 (iii) below). Therefore, the point $\tilde{u}^{k+1} = u^k - \gamma\sigma_k g^k$ is the projection of the point u^k onto the hyperplane

$$H_k(\gamma) = \{v \in R^n \mid \langle g^k, v - u^k \rangle = -\gamma\sigma_k \|g^k\|^2\}.$$

Clearly, $H_k(1)$ separates u^k and U^d. Hence, the distance from $\tilde{u}^{k+1}$ to each point of U^d cannot increase when $\gamma \in (0,2)$ and that from u^{k+1} does so due to the properties of P_k. Thus, our method follows the general combined relaxation framework, hence it enjoys convergence properties similar to those of methods from Chapter 1. More precisely, Method 2.2 follows the slightly modified Basic Scheme II from Section 1.6.

Note that the indices t and $k(l)$, the index sets $I_{k,l}$ and K_l, and the sequence $\{y^l\}$ are introduced only for simplicity of further exposition, but, generally speaking, can be dropped. We will call one increase of the index t an inner step, so that the number of inner steps gives the number of computations of elements from $Q(\cdot)$ at the corresponding points.

As to the auxiliary procedure in Step 1, it is a modification of an iteration of the simple relaxation subgradient method (see [95, 97]). For the convenience of the reader, we describe this method in detail in Section 4.3.

2.3.2 Properties of Auxiliary Mappings

Let us define the mapping $P : R^n \to \Pi(R^n)$ by

$$P(u) = \begin{cases} G(u) & \text{if } h(u) < 0, \\ \text{conv}\{G(u) \bigcup \partial h(u)\} & \text{if } h(u) = 0, \\ \partial h(u) & \text{if } h(u) > 0. \end{cases} \qquad (2.48)$$

We now obtain some properties of the mappings Q and P, which are intended to be used for proving convergence of our method. In particular, we will show that GVI (2.1) is equivalent to the multivalued inclusion

$$0 \in P(u^*). \qquad (2.49)$$

We denote by S^* the solution set of problem (2.49).

We first establish continuity properties for Q and P.

Proposition 2.3.1. *The mappings Q and P, being defined by (2.43) and (2.48), are K-mappings.*

Proof. On account of Proposition 2.1.8 (iii), ∂h is a K-mapping. It follows that Q and P have nonempty convex and compact values. Using properties of u.s.c. mappings (e.g., see [210, Theorem 5.1.4]), we conclude that $\text{conv}\{G \bigcup \partial h\}$ is u.s.c. and that so are Q and P. $\qquad \Box$

Lemma 2.3.1. *Let W and Y be convex compact sets in R^n, $0 \notin Y$. Then the relation $0 \in \text{conv}\{W \cup Y\}$ is equivalent to $0 \in W + \text{cone}Y$.*

Proof. Suppose $0 \in W + \text{cone}Y$, then there are $a \in W$, $b \in Y$ and $\lambda \geq 0$, such that $a + \lambda b = 0$, or equivalently, $(1 + \lambda)^{-1}a + \lambda(1 + \lambda)^{-1}b = 0$, i.e., $0 \in \text{conv}\{W \cup Y\}$. Conversely, let $0 \in \text{conv}\{W \cup Y\}$, then there are $a \in W$, $b \in Y$ and $\mu \in [0,1]$ such that $\mu a + (1 - \mu)b = 0$. If $\mu = 0$, then $b = 0$, which contradicts the fact $0 \notin Y$. Hence, we must have $\mu \in (0,1]$ and $0 = a + \mu^{-1}(1 + \mu)b \in W + \text{cone}Y$, as desired. $\qquad \Box$

Lemma 2.3.2. *If $u \in U$, then*

$$N(U,u) = \begin{cases} 0 & \text{if } h(u) < 0, \\ \text{cone}\partial h(u) & \text{if } h(u) = 0. \end{cases}$$

Proof. If $h(u) < 0$, then $u \in \text{int}U$ and, by definition, $N(U,u) = \{0\}$. If $h(u) = 0$, then u cannot be a minimizer for h. Since $\partial h(u)$ is compact, we must have $N(U,u) = \text{cone}\partial h(u)$ due to [185, Theorem 23.7]. $\qquad \Box$

Theorem 2.3.1. *It holds that*

$$U^* = S^*.$$

Proof. It follows from the Slater condition that $0 \notin \partial h(u)$ if $h(u) \geq 0$. So, if (2.49) holds, then $u^* \in U$. By setting $W = G(u^*)$ and $Y = \partial h(u^*)$ in Lemma 2.3.1 we see that (2.49) is equivalent to

$$u^* \in U, \quad 0 \in \begin{cases} G(u^*) & \text{if } h(u^*) < 0, \\ G(u^*) + \text{cone}\partial h(u^*) & \text{if } h(u^*) = 0. \end{cases}$$

Now, it follows from Lemma 2.3.2 that relations (2.49) and (2.1) are equivalent. □

Note that the assertion of Theorem 2.3.1 is similar to optimality conditions in nonsmooth optimization; e.g., see [6, 44, 105].

2.3.3 Convergence

In order to obtain a convergence result for Method 2.2, we need to investigate properties of its auxiliary procedure in Step 1. It is a modification of an iteration of the simple relaxation subgradient method (see Section 4.3 for details). This procedure can be completed due to two different conditions. In the first case, the procedure generates a descent direction for the distance function $\text{dist}(u) = \|u - u^*\|$ where u is a given point and u^* is any point of U^d. This is called a descent step. Otherwise, if the descent step does not occur, we will show that the procedure generates a sequence $\{p^i\}$ such that $\|p^i\| \searrow 0$, p^i belongs to the convex hull of vectors from $P(w^s)$ where $\|w^s - u\| \leq \varepsilon_l$, $s = 0, 1, \ldots, i$. It follows that the stopping test (2.45) at Step 1.2 will be passed in a finite number of iterations and, taking into account Theorem 2.3.1, one can conclude that the point u approximates a point of U^* within given positive tolerances ε_l and η_l. This is called a null step.

Lemma 2.3.3. *Let k and l be fixed. Then:*
(i)

$$p^i \in \text{conv}\{q^0, \ldots, q^i\} \tag{2.50}$$

and
(ii)

$$\|p^i\| \leq D_{k,l}/((1 - \theta)\sqrt{i + 1}), \tag{2.51}$$

where

$$D_{k,l} = \max_{i \in I_{k,l}} \|q^i\|. \tag{2.52}$$

Proof. Assertions (i) and (ii) follows from Lemmas 4.3.1 and 4.3.2, respectively. □

We are now ready to establish the basic properties of Procedure P.

Lemma 2.3.4. *Let k and l be fixed. Then:*
(i) The number of inner steps is finite.
(ii) It holds that

$$p^i \in \text{conv} \bigcup_{\|v-u^k\| \le \varepsilon_l} P(v) \quad \text{for} \quad i = 0, 1, \dots$$

(iii) In the case of a descent step, for every $u^* \in U^d$, we have

$$\langle g^k, u^k - u^* \rangle \ge \langle g^k, u^k - v^k \rangle > \theta \varepsilon_l \eta_l.$$

Proof. Part (i) follows from Lemma 2.3.3 (ii), Proposition 2.3.1 and from the stopping test (2.45). Also, since $Q(u) \subseteq P(u)$, part (ii) follows from Lemma 2.3.3 (i). Next, let the auxiliary procedure terminate at Step 1.3. Then we must have $v^k = w^{i+1}$. If $w^{i+1} \notin U$, then $q^{i+1} \in \partial h(w^{i+1})$ and

$$\varepsilon_l \langle q^{i+1}, p^i / \|p^i\| \rangle = \langle q^{i+1}, u^k - w^{i+1} \rangle \le h(u^k) - h(w^{i+1}) < 0,$$

which contradicts (2.46). Therefore, $w^{i+1} \in U$ and $q^{i+1} \in G(w^{i+1})$. Take any $u^* \in U^d$. Then, taking into account (2.46) and the stopping test (2.45) in Step 1.2, we obtain

$$
\begin{aligned}
\langle g^k, u^k - u^* \rangle &= \langle g^k, u^k - v^k \rangle + \langle g^k, v^k - u^* \rangle \\
&\ge \langle g^k, u^k - v^k \rangle = \langle q^{i+1}, u^k - w^{i+1} \rangle \\
&= \varepsilon_l \langle q^{i+1}, p^i / \|p^i\| \rangle \ge \theta \varepsilon_l \|p^i\| > \theta \varepsilon_l \eta_l.
\end{aligned}
$$

This proves part (iii). $\qquad\qquad\qquad\qquad\qquad\qquad\qquad\qquad\qquad\qquad\square$

We now obtain the key properties of the main iteration sequence $\{u^k\}$.

Lemma 2.3.5. (i) For every $u^* \in U^d$, we have

$$\|u^{k+1} - u^*\|^2 \le \|u^k - u^*\|^2 - \gamma_{j(k)}(2 - \gamma_{j(k)})(\sigma_k \|g^k\|)^2 \qquad (2.53)$$

for $k = 0, 1, \dots$
  (ii) Sequences $\{u^k\}$ and $\{v^k\}$ are bounded.
  (iii)

$$\sum_{k=0}^{\infty} \gamma_{j(k)}(2 - \gamma_{j(k)})(\sigma_k \|g^k\|)^2 < \infty.$$

(iv) If $\sigma_k \|g^k\| \ge \tilde{\sigma} > 0$ as $l \le s < \infty$, the sequence $\{y^l\}$ is infinite.
  (v) For each limit point u^* of $\{u^k\}$ such that $u^* \in U^d$, we have

$$\lim_{k \to \infty} u^k = u^*.$$

Proof. On account of Lemma 2.3.4, part (i) is argued similarly to the proof of Lemma 1.2.3. Parts (iii) and (v) follow directly from (i). Part (iv) is argued similarly to the proof of Lemma 1.6.1(iii). Part (ii) follows from (i) and from the construction of Method 2.2. $\qquad\qquad\qquad\qquad\qquad\qquad\qquad\square$

We now establish a convergence result for Method 2.2.

Theorem 2.3.2. *Let a sequence $\{u^k\}$ be generated by Method 2.2 and let $\{\varepsilon_l\}$ and $\{\eta_l\}$ satisfy the following relations:*

$$\{\varepsilon_l\} \searrow 0, \{\eta_l\} \searrow 0. \tag{2.54}$$

Then:

   (i) The number of inner steps at each iteration is finite.
   (ii) There exists a limit point u^* of $\{u^k\}$ which lies in U^*.
   (iii) If, in addition,

$$U^* = U^d, \tag{2.55}$$

we have

$$\lim_{k\to\infty} u^k = u^* \in U^*.$$

Proof. Part (i) follows from Lemmas 2.3.4 (i) and 2.3.5 (ii). Next, if $l \leq s < \infty$, then there exists a number $\bar{k}$ such that all the iterations of the auxiliary procedure terminate at Step 1.3 when $k \geq \bar{k}$. By Lemma 2.3.5 (ii), $\{v^k\}$ is bounded, hence so is $\{g^k\}$, i.e.,

$$\|g^k\| \leq C_2 < \infty, \quad k = 0, 1, \dots$$

Taking into account Lemma 2.3.4 (iii), we now have

$$\sigma_k \|g^k\| = \langle g^k, u^k - v^k \rangle / \|g^k\| \geq \theta \varepsilon_s \eta_s / C_2 = \tilde{\sigma} > 0,$$

therefore, by Lemma 2.3.5 (iv), $\{y^l\}$ is infinite. Moreover, $\{y^l\}$ is bounded due to Lemma 2.3.5 (ii), hence, there exists a subsequence of $\{y^l\}$ that converges to some point $u^* \in U$. Without loss of generality we suppose that

$$\lim_{l\to\infty} y^l = u^*.$$

On the other hand, according to Lemma 2.3.4 (ii) and the construction of the method, there exists an element $p^{(l)}$ such that

$$\|p^{(l)}\| \leq \eta_l, p^{(l)} \in \text{conv} \bigcup_{\|v-y^l\|\leq\varepsilon_l} P(v)$$

for $l = 1, 2, \dots$ Since P is u.s.c. due to Proposition 2.3.1, these properties together with (2.54) yield $u^* \in S^*$ or equivalently, $u^* \in U^*$ due to Theorem 2.2.4. Hence, the assertion of (ii) is true. Part (iii) now follows from (2.55) and Lemma 2.3.5 (v). The proof is complete. $\square$

2.3.4 Complexity Estimates

In this section, we give some complexity estimates for Method 2.2. As Method 2.2 has a two-level structure with each iteration containing a finite number of inner steps, it is more suitable to derive its complexity estimate, which gives the total amount of work of the method.

First we consider the following additional hypothesis:

(HC2.3)

(a)

$$\gamma_j = \gamma \in (0,2) \quad for \; k = 0,1,\dots; \tag{2.56}$$

(b) *there exists* $u^* \in U^*$ *such that*

$$for \; every \; u \in U \; and \; for \; every \; g \in G(u), \atop \langle g, u - u^* \rangle \geq \mu \|u - u^*\|^\varepsilon, \tag{2.57}$$

where $\varepsilon > 0$ *and* $\mu > 0$.

It is easy to see that (2.57) slightly generalizes (2.2) in the case where $\mu = 0$. Next, if $\varepsilon = 1$, then (2.57) can be viewed as a modification of Assumption (A6). Besides, if G is strongly monotone with constant μ, then (2.57) holds with $\varepsilon = 2$. It should be also noted that condition (2.57) implies that $U^* = \{u^*\}$, i.e. U^* contains at most one point.

In this section, we use the distance to u^* as an accuracy function for our method, i.e., our approach is slightly different from those in [155, 133, 143]. More precisely, given a starting point u^0 and a number $\delta > 0$, we define the complexity of the method, denoted by $N(\delta)$, as the total number of inner steps t which ensures finding a point $\bar{u} \in U$ such that

$$\|\bar{u} - u^*\| / \|u^0 - u^*\| \leq \delta.$$

Recall that each inner step includes one computation of an element from $Q(\cdot)$ at the corresponding point. Therefore, since the computational expense per inner step can easily be evaluated for each specific problem under examination, this estimate in fact gives the total amount of work.

We now proceed to obtain an upper bound for $N(\delta)$. First we establish several auxiliary properties.

Lemma 2.3.6. *Let a sequence* $\{u^k\}$ *be generated by Method 2.2 and let* $(HC2.3)$ *hold. Then*

$$\|g^k\| \leq C' < \infty \; and \; D_{k,l} \leq C' < \infty. \tag{2.58}$$

Proof. By Lemma 2.3.5 (ii), $\{u^k\}$ and $\{v^k\}$ are bounded, hence, so are $\{g^k\}$ and $D_{k,l}$ due to (2.52) and the properties of G and ∂h. $\qquad\square$

In what follows, for simplicity, we set $D_l = D_{k(l),l}$. Fix a point $\bar{y}$ of U such that $h(\bar{y}) < 0$.

Lemma 2.3.7. *Let a sequence* $\{u^k\}$ *be generated by Method 2.2 and let* $(HC2.3)$ *hold. For each* $l = 1, 2, \dots$, *there exists a number* $\tau_l \in [0,1]$ *such that the following holds:*

(i) if (2.57) holds with $\varepsilon = 1$, *then*

$$\eta_l\|u^* - y^l\| \geq \tau_l\mu\|u^* - y^l\| - \varepsilon_l(\tau_l\mu + D_l);$$
(2.59)

(ii) if G is monotone on U, then

$$\eta_l\|\bar{y} - y^l\| + \varepsilon_l D_l \geq \tau_l(\langle \bar{g}, y^l - \bar{y}\rangle - \varepsilon_l\|\bar{g}\|) - (1 - \tau_l)h(\bar{y}),$$
(2.60)

where $\bar{g} \in G(\bar{y})$.

Proof. Fix any $l = 1, 2, \ldots$ Let us consider the $k(l)$th iteration of Method 2.2. By (2.50), for each s, we have

$$p^s = \sum_{i=0}^{s}\beta_i q^i, \quad \sum_{i=0}^{s}\beta_i = 1; \quad \beta_i \geq 0, \quad \text{for } i = 0, \ldots, s.$$
(2.61)

By definition, at the $k(l)$th iteration, the auxiliary procedure produces a null step, i.e., there exists a number $\bar{s}$ such that

$$\|p^{\bar{s}}\| \leq \eta_l$$
(2.62)

(see (2.45)). For simplicity, we omit the indices of the set $I_{k(l),l}$ in this proof, i.e., let

$$I = \{0, \ldots, \bar{s}\}, \quad \bar{I} = \{i \in I \mid w^i \in U\}, \quad \tilde{I} = \{i \in I \mid w^i \notin U\}.$$

Then

$$q^i \in \begin{cases} G(w^i) & \text{if } i \in \bar{I}, \\ \partial h(w^i) & \text{if } i \in \tilde{I}. \end{cases}$$
(2.63)

Set

$$\tau_l = \sum_{i \in \bar{I}}\beta_i.$$
(2.64)

It is obvious that $\tau_l \in [0, 1]$. Observe also that

$$\|w^i - y^l\| \leq \varepsilon_l \quad \text{for } i \in I.$$
(2.65)

In the case of (i), we have

$$\langle q^i, w^i - u^*\rangle \geq \mu\|w^i - u^*\| \quad \text{for } i \in \bar{I}$$

and

$$\langle q^i, w^i - u^*\rangle \geq 0 \quad \text{for } i \in \tilde{I};$$

where $\{u^*\} = U^*$. Applying (2.65) gives

$$\begin{aligned} \langle q^i, y^l - u^*\rangle &\geq \langle q^i, y^l - w^i\rangle + \mu\|w^i - u^*\| \\ &\geq -\varepsilon_l\|q^i\| + \mu(\|u^* - y^l\| - \|y^l - w^i\|) \\ &\geq -\varepsilon_l\|q^i\| + \mu(\|u^* - y^l\| - \varepsilon_l) \end{aligned}$$

for $i \in \bar{I}$ and

$$\langle q^i, y^l - u^* \rangle \geq \langle q^i, y^l - w^i \rangle \geq -\varepsilon_l \|q^i\|$$

for $i \in \tilde{I}$. Summing these inequalities multiplied by β_i, over $i \in I$ and taking into account (2.52), (2.62), and (2.64), we obtain

$$\eta_l \|y^l - u^*\| \geq \langle p^s, y^l - u^* \rangle \geq -\varepsilon_l D_l + \tau_l \mu(\|u^* - y^l\| - \varepsilon_l),$$

i.e. (2.59) holds.

In the case of (ii), by the monotonicity of G and by (2.63), for every $\bar{g} \in G(\bar{y})$, we have

$$\langle q^i, w^i - \bar{y} \rangle \geq \langle \bar{g}, w^i - \bar{y} \rangle \quad \text{for } i \in \tilde{I}.$$

Also, by (2.63) and the subgradient property,

$$\langle q^i, w^i - \bar{y} \rangle \geq h(w^i) - h(\bar{y}) \geq -h(\bar{y}) \quad \text{for } i \in \tilde{I}.$$

Applying (2.65) gives

$$\begin{aligned}\langle q^i, y^l - \bar{y} \rangle &\geq \langle q^i, y^l - w^i \rangle + \langle \bar{g}, y^l - \bar{y} \rangle + \langle \bar{g}, w^i - y^l \rangle \\ &\geq -\varepsilon_l \|q^i\| + \langle \bar{g}, y^l - \bar{y} \rangle - \varepsilon_l \|\bar{g}\|\end{aligned}$$

for $i \in \bar{I}$ and, respectively,

$$\langle q^i, y^l - \bar{y} \rangle \geq \langle q^i, y^l - w^i \rangle - h(\bar{y}) \geq -\varepsilon_l \|q^i\| - h(\bar{y})$$

for $i \in \tilde{I}$. Summing these inequalities multiplied by β_i, over $i \in I$ and taking into account (2.52), (2.62), and (2.64), we obtain

$$\eta_l \|y^l - \bar{y}\| \geq \langle p^s, y^l - \bar{y} \rangle \geq -\varepsilon_l D_l + \tau_l(\langle \bar{g}, y^l - \bar{y} \rangle - \varepsilon_l \|\bar{g}\|) - (1 - \tau_l)h(\bar{y}),$$

i.e. (2.60) holds and the proof is complete. $\qquad\square$

We now show that Method 2.2 attains a logarithmic complexity estimate. Denote by $l(\delta)$ the maximal value of the index l in Method 2.2 such that

$$\|y^l - u^*\| \geq \delta.$$

Taking into account the definition of the inner step, we conclude that the following bound is true:

$$N(\delta) \leq \sum_{l=1}^{l(\delta)+1} \sum_{k \in K_l} (|I_{k,l}| - 1). \qquad (2.66)$$

We proceed to estimate the right-hand side of (2.66). We first establish a complexity estimate for the set-valued case.

Theorem 2.3.3. *Suppose G is monotone and Hypothesis (HC2.3) holds with $\varepsilon = 1$. Let a sequence $\{u^k\}$ be generated by Method 2.2 where*

$$\varepsilon_l = \nu^l \varepsilon', \eta_l = \eta', l = 0, 1, \ldots; \quad \nu \in (0, 1). \tag{2.67}$$

Then, there exist some constants $\bar{\varepsilon} > 0$ and $\bar{\eta} > 0$ such that

$$N(\delta) \leq B_1 \nu^{-2}(\ln(B_0/\delta)/\ln \nu^{-1} + 1), \tag{2.68}$$

where $0 < B_0, B_1 < \infty$, whenever $0 < \varepsilon' \leq \bar{\varepsilon}$ and $0 < \eta' \leq \bar{\eta}$, B_0 and B_1 being independent of ν.

Proof. We first show that $\{y^l\}$ is infinite under the conditions of the present theorem. Indeed, otherwise there exist numbers $\bar{l}$ and $\bar{k}$ such that $k \in K_{\bar{l}}$ when $k \geq \bar{k}$. Now, using (2.46) and (2.58) gives

$$\sigma_k \|g^k\| = \langle g^k, u^k - v^k \rangle / \|g^k\| \geq \theta \varepsilon_{\bar{l}} \eta_{\bar{l}}/C',$$

and the result now follows from Lemma 2.3.5 (iv).

Next, by (2.51), (2.58), and (2.67), the number of inner steps for any fixed k and l does not exceed the value

$$C = (C'/((1 - \theta)\eta'))^2 \geq |I_{k,l}| - 1. \tag{2.69}$$

We now proceed to show that

$$\|y^l - u^*\| \leq B_0 \nu^l, l = 0, 1, \ldots, 0 \leq B_0 < \infty. \tag{2.70}$$

By Lemma 2.3.5 (ii), $\{u^k\}$ is bounded, hence

$$\|\bar{y} - y^l\| \leq B_3 < \infty, \quad l = 1, 2, \ldots$$

Therefore, by (2.58) and (2.60) we have

$$\eta_l(-h(\bar{y}) + (\varepsilon_l + B_3)\|\bar{g}\|) \geq -h(\bar{y}) - \eta_l B_3 - \varepsilon_l C'.$$

Letting

$$\eta_l \leq \tilde{\eta} = -h(\bar{y})/(4B_3), \quad \varepsilon' \leq \bar{\varepsilon} = -h(\bar{y})/(4C'),$$

we have $\varepsilon_l \leq \bar{\varepsilon}$ and

$$\eta_l \geq \tau' = -h(\bar{y})/(2(-h(\bar{y}) + (\bar{\varepsilon} + B_3)\|\bar{g}\|)) > 0$$

for $l = 1, 2, \ldots$ By using (2.59) and the fact that $\eta_l \in [0, 1]$, we have

$$(\eta_l \mu - \eta_l)\|y^l - u^*\| \leq \varepsilon_l(\mu + C').$$

Therefore, letting

$$\eta_l = \eta' \leq \bar{\eta} = \min\{\tilde{\eta}, \tau' \mu/2\},$$

we see that (2.70) holds with

$$B_0 = \max\{\|y^0 - u^*\|, \bar{\varepsilon}(\mu + C')/(\tau'\mu - \bar{\eta})\}.$$

Next, take any k such that $k(l) < k < k(l+1)$ and set $d = k - k(l)$. Then, $\sigma_k > 0$ and using (2.46), (2.58), and (2.67) gives

$$\sigma_k \|g^k\| \geq \theta\varepsilon_{l+1}\eta_{l+1}/\|g^k\| \geq \theta\varepsilon_{l+1}\eta'/C'.$$

Combining this inequality with (2.56), (2.53), and (2.67) gives

$$0 \leq \|u^{k+1} - u^*\|^2 \leq \|u^{k(l)} - u^*\|^2 - \gamma(2 - \gamma)\sum_{s=k(l)}^{k}(\sigma_s\|g^s\|)^2$$

$$\leq \|y^l - u^*\|^2 - d\gamma(2 - \gamma)(\theta\varepsilon_{l+1}\eta'/C')^2.$$

Applying (2.70) in this inequality now gives

$$|K_{l+1}| \leq (B_0 C'/(\theta\varepsilon'\eta'))^2/(\nu^2\gamma(2 - \gamma)) = B_1'\nu^{-2}. \tag{2.71}$$

On the other hand, from (2.70) we have

$$l(\delta) \leq \ln(B_0/\delta)/\ln\nu^{-1}. \tag{2.72}$$

Applying this inequality together with (2.69) and (2.71) in (2.66) yields

$$N(\delta) \leq CB_1'\nu^{-2}(\ln(B_0/\delta)/\ln\nu^{-1} + 1),$$

i.e., relation (2.68) holds with $B_1 = CB_1'$.   □

It should be noted that the assertion of Theorem 2.3.3 remains valid without the additional monotonicity assumption on G if $U = R^n$. In fact, on account of (2.61) and (2.64) we then have $\tau_l = 1$ and (2.70) follows directly from (2.59) and (2.67).

Thus, our method attains a logarithmic complexity estimate, which corresponds to a linear rate of convergence with respect to inner steps. We now establish a similar upper bound for $N(\delta)$ in the single-valued case.

Theorem 2.3.4. *Suppose that $U = R^n$ and that Hypothesis (HC2.3) holds with $\varepsilon = 2$. Suppose also that G is Lipschitz continuous with constant L. Let a sequence $\{u^k\}$ be generated by Method 2.2 where*

$$\varepsilon_l = \nu^l\varepsilon', \eta_l = \nu^l\eta', l = 0, 1, \ldots; \varepsilon' > 0, \eta' > 0; \quad \nu \in (0, 1). \tag{2.73}$$

Then,

$$N(\delta) \leq B_1\nu^{-6}(\ln(B_0/\delta)/\ln\nu^{-1} + 1), \tag{2.74}$$

where $0 < B_0, B_1 < \infty$, B_0 and B_1 being independent of ν.

Proof. By using the same argument as that in the proof of Theorem 2.3.2 we deduce that $\{y^l\}$ is infinite. We now proceed to obtain relations similar to (2.69) – (2.72). Let us consider the kth iteration of Method 2.2, $k(l) < k < k(l+1)$. Due to the conditions of the present theorem, $U^* = \{u^*\}$. Then we must have

$$\|u^k - u^*\| \le \|y^l - u^*\| \le \|G(y^l)\|/\mu \tag{2.75}$$

and

$$\mu\|u^k - u^*\| \le \|G(u^k)\| \le L\|u^k - u^*\| \tag{2.76}$$

due to (2.53), $(HC2.3)$ and the fact that $G(u^*) = 0$. Besides, since

$$\|u^k - w^i\| \le \varepsilon_{l+1}$$

for $i \in I_{k,l+1}$, we see that

$$\|q^i - G(u^k)\| \le L\varepsilon_{l+1} \tag{2.77}$$

and, on account on (2.61),

$$\|p^s - G(u^k)\| \le L\varepsilon_{l+1}, \quad \text{for } s \in I_{k,l+1}. \tag{2.78}$$

It follows from (2.75)–(2.77) that

$$\begin{aligned} D_{k,l} &\le \|G(u^k)\| + L\varepsilon_{l+1} \le L(\|y^l - u^*\| + \varepsilon_{l+1}) \\ &\le L(\|G(y^l)\|/\mu + \varepsilon_{l+1}). \end{aligned} \tag{2.79}$$

On the other hand, at the $k(l)$th iteration, there exists a number $\bar{s}$ such that $\|p^{\bar{s}}\| \le \eta_l$ (see (2.62)) and $\|p^{\bar{s}} - G(y^l)\| \le L\varepsilon_l$ due to $\|w^i - y^l\| \le \varepsilon_l$. Hence,

$$\|G(y^l)\| \le \|p^{\bar{s}}\| + L\varepsilon_l \le \eta_l + L\varepsilon_l. \tag{2.80}$$

Using this inequality in (2.79) gives

$$\begin{aligned} D_{k,l} &\le L((\eta_l + L\varepsilon_l)/\mu + \varepsilon_{l+1}) \\ &\le L((\eta' + L\varepsilon')/(\mu\nu) + \varepsilon')\nu^{l+1}. \end{aligned}$$

Combining this inequality with (2.51), (2.73) and taking into account the stopping criterion (2.45), we obtain

$$|I_{k,l}| - 1 \le C''\nu^{-2} \tag{2.81}$$

where

$$C'' = (L((\eta' + L\varepsilon')/\mu + \varepsilon')/((1-\theta)\eta'))^2.$$

Next, from (2.75) and (2.80) we obtain

$$\|y^l - u^*\| \le (\eta' + L\varepsilon')\nu^l/\mu,$$

i.e., (2.70) holds with

$$B_0 = \max\{\|y^0 - u^*\|, (\eta' + L\varepsilon')/\mu\}.$$

Further, using (2.46), (2.70), (2.73), and (2.75)–(2.79) gives

$$
\begin{aligned}
\sigma_k \|g^k\| &= \langle g^k, u^k - v^k\rangle / \|g^k\| \geq \theta \varepsilon_{l+1} \eta_{l+1} / \|g^k\| \\
&\geq \theta \varepsilon_{l+1} \eta_{l+1} / (\|G(u^k)\| + L\varepsilon_{l+1}) \\
&\geq \theta \varepsilon_{l+1} \eta_{l+1} / (L(\|y^l - u^*\| + \varepsilon_{l+1})) \\
&\geq \theta \varepsilon_{l+1} \nu / (L(B_0 + \varepsilon')) = B_4' \nu^{l+2}.
\end{aligned}
$$

Applying this inequality in (2.53) and taking into account (2.56) and (2.73), we have

$$
\begin{aligned}
0 \leq \|u^{k+1} - u^*\|^2 &\leq \|u^{k(l)} - u^*\|^2 - \gamma(2-\gamma) \sum_{s=k(l)}^{k} (\sigma_s \|g^s\|)^2 \\
&\leq \|y^l - u^*\|^2 - d\gamma(2-\gamma)(B_4' \nu^{l+2})^2
\end{aligned}
$$

for $k = k(l) + d$. On account of (2.70) we now have (2.72) and

$$|K_{l+1}| \leq \nu^{-4}(B_0/B_4')^2/(\gamma(2-\gamma)) = B_4 \nu^{-4}.$$

Applying this inequality together with (2.72) and (2.81) in (2.66) gives

$$N(\delta) \leq C'' B_4 \nu^{-6} (\ln(B_0/\delta)/\ln \nu^{-1} + 1),$$

i.e., relation (2.74) holds with $B_1 = C'' B_4$. The proof is complete. $\square$

2.3.5 Modifications

Method 2.2 can be adjusted to solve GVI (2.1) in the general case where the feasible set U is not associated to some function h. Namely, we can consider the following assumptions:

Hypothesis (HG2.3)

(a) U is a nonempty, closed and convex subset of R^n such that $\mathrm{int}U \neq \emptyset$;
(b) $G : U \to \Pi(R^n)$ is a K-mapping;
(c) $U^d \neq \emptyset$.

Fix $\rho > 0$ and set

$$
\begin{aligned}
T_\rho(u) &= N(U, u) \bigcap S(0, \rho), \\
D(u) &= \mathrm{conv} S_\rho(u)
\end{aligned}
\tag{2.82}
$$

for every $u \in R^n$. We can now replace the mappings Q and P with the following mappings $\tilde{Q} : R^n \to \Pi(R^n)$ and $\tilde{P} : R^n \to \Pi(R^n)$, respectively, which are defined by

$$\tilde{Q}(u) = \begin{cases} G(u) & \text{if } u \in U, \\ D(u) & \text{if } u \notin U; \end{cases} \tag{2.83}$$

and

$$\tilde{P}(u) = \begin{cases} G(u) & \text{if } u \in \text{int}U, \\ \text{conv}\{G(u) \cup D(u)\} & \text{if } u \in U \backslash \text{int}U, \\ D(u) & \text{if } u \notin U. \end{cases} \tag{2.84}$$

We first obtain an analogue of Proposition 2.3.1.

Proposition 2.3.2. *The mappings D, $\tilde{Q}$ and $\tilde{P}$ being defined by (2.82) – (2.84) are K-mappings.*

Proof. It is clear that $N(U, \cdot)$ is a closed mapping; e.g. see [44, Chapter 1, Section 13.1 and Chapter 4, Section 10.1]. Therefore, so is T_ρ, but $T_\rho(u)$ is compact for each $u \in R^n$ and $T_\rho(Z)$ is compact for each compact set $Z \subseteq R^n$. By Proposition 2.1.1 (ii), T_ρ is u.s.c. Using the properties of u.s.c. mappings (e.g. see [210, Theorem 5.1.4]), we conclude that D, $\tilde{Q}$ and $\tilde{P}$ are also u.s.c. $\square$

We denote by $\tilde{S}^*$ the whole set of stationary points of $\tilde{P}$, i.e., $u^* \in \tilde{S}^*$ if and only if $0 \in \tilde{P}(u^*)$.

Theorem 2.3.5. *It holds that*

$$U^* = \tilde{S}^*.$$

Proof. Since $\text{int}U \neq \emptyset$, there exist a point $\bar{y} \in U$ and a number $\varepsilon > 0$ such that $B(\bar{y}, \varepsilon) \subseteq U$ due to the closedness of U. It follows that, for every $q \in T_\rho(u^*)$ and for every $u^* \notin \text{int}U$,

$$\langle q, u - u^* \rangle \leq 0 \quad \forall u \in B(\bar{y}, \varepsilon),$$

whence we have

$$\langle q, \bar{y} - u^* \rangle \leq -\varepsilon\rho.$$

Therefore,

$$\|q\| \geq \varepsilon\rho / \|u^* - \bar{y}\| > 0 \quad \forall q \in D(u^*).$$

Thus, $0 \notin D(u^*)$ when $u^* \notin \text{int}U$. It follows that $\tilde{S}^* \subseteq U$. Moreover, if $u^* \in U \backslash \text{int}U$ and $0 \in \tilde{P}(u^*)$, then $0 \notin D(u^*)$ and $N(U, u^*) = \text{cone}D(u^*)$. By Lemma 2.3.1 with $W = G(u^*)$ and $Y = D(u^*)$, the relation $0 \in \tilde{P}(u^*)$ is then equivalent to

$$0 \in G(u^*) + N(U, u^*).$$

The same is true in the case where $u^* \in \mathrm{int}U$, since we then have $N(U, u^*) = \{0\}$. The proof is complete.   □

Thus, we can substitute Q with $\tilde{Q}$ in Method 2.2 and P with $\tilde{P}$ in the corresponding proofs. Then, it is easy to see that all the results of Theorems 2.3.2 and 2.3.3 remain valid, after little modifications.

Next, the simplest rule (2.47) in Method 2.2 can be replaced by one of the following:

$$p^{i+1} = \mathrm{Nr}\ \mathrm{conv}\{q^0, \ldots, q^{i+1}\}, \tag{2.85}$$

or

$$p^{i+1} = \mathrm{Nr}\ \mathrm{conv}\{p^i, q^{i+1}, S_i\}, \tag{2.86}$$

where $S_i \subseteq \mathrm{conv}\{q^0, \ldots, q^i\}$. From the definition it follows that the assertion of Lemma 2.3.3 remains valid after this modification and that the proofs of the other results are not changed. Thus, one can make use of rule (2.85) or (2.86) to accelerate the convergence of the method, but it should be noted that they need the additional storage and work per inner step.

2.4 CR Method for Multivalued Inclusions

In this section, we again consider GVI (2.1) under the blanket assumptions of Section 2.3. To solve this problem, we propose to apply Method 2.2 to finding stationary points either of the mapping P being defined in (2.48) or the mapping $\tilde{P}$ being defined in (2.84), respectively. Such a method need not include feasible quasi-nonexpansive mappings.

2.4.1 An Inexact CR Method for Multivalued Inclusions

We first consider GVI (2.1) under Hypothesis (H2.3), i.e., in the case where the feasible set U is defined by a convex function h. Then, according to Theorem 2.3.1, the solution set U^* of GVI (2.1) and the set of stationary points S^* of the mapping $P : R^n \to \Pi(R^n)$ being defined in (2.48) coincide. Moreover, P is then a K-mapping due to Proposition 2.3.1. Hence, the initial GVI (2.1) can be replaced by the multivalued inclusion (2.49). However, in order to apply Method 2.2 to problem (2.49) we have to show that its dual problem is solvable. Namely, let us consider the problem of finding a point u^* of R^n such that

$$\forall u \in R^n, \quad \forall t \in P(u), \quad \langle t, u - u^* \rangle \geq 0, \qquad (2.87)$$

which can be viewed as the dual problem to (2.49). We denote by $S^*_{(d)}$ the solution set of this problem. From Proposition 2.1.3 we obtain the following.

Lemma 2.4.1. (i) $S^*_{(d)}$ is convex and closed.

(ii) $S^*_{(d)} \subseteq S^*$.

(iii) If P is pseudomonotone, then $S^*_{(d)} = S^*$.

Note that P need not be pseudomonotone in general. Nevertheless, in addition to Theorem 2.3.1, it is useful to obtain the equivalence result for problems (2.2) and (2.87).

Proposition 2.4.1. $U^d = S^*_{(d)}$.

Proof. Take any points $u^* \in U^d$ and $u \in R^n$. If $h(u) \leq 0$ then for each $t \in G(u)$, we have $\langle t, u - u^* \rangle \geq 0$ because of (2.2). If $h(u) \geq 0$, then, by definition,

$$0 \geq h(u^*) - h(u) \geq \langle t, u^* - u \rangle \quad \forall t \in \partial h(u).$$

On account of (2.48), we obtain $u^* \in S^*_{(d)}$. Hence, $U^d \subseteq S^*_d$. Conversely, by Lemma 2.4.1 (ii) and Theorem 2.3.1, we have $S^*_{(d)} \subseteq S^* = U^* \subseteq U$. Assume for contradiction that there exists an element $z^* \in S^*_{(d)} \backslash U^d$. Then there exist a point $u \in U$ and an element $g \in G(u)$ such that $\langle g, u - z^* \rangle < 0$. But $G(u) \subseteq P(u)$, hence $z^* \notin S^*_{(d)}$, a contradiction. The proof is complete. $\qquad \square$

Thus, under Hypothesis $(H2.3)$, P is a K-mapping and $S^*_{(d)}$ is nonempty. Since the multivalued inclusion (2.49) is a particular case of GVI, we can apply Method 2.2 to problem (2.49). In such a way, we obtain a solution of the initial problem (2.1) as well. Moreover, we consider an inexact version of Method 2.2, which admits inexact computations of elements of $P(u^k)$. Such a technique allows one to take into account various computation errors and to partially "smooth" the initial problem since any element of $P(u^k)$ can be now replaced with the convex hull of elements being computed near u^k. For each point $u \in R^n$, set

$$F_\delta(u) = \text{conv} \bigcup_{u \in B(v,\delta)} P(u).$$

Method 2.3. *Step 0 (Initialization)*: Choose a point $u^0 \in R^n$, a sequence $\{\gamma_j\}$ satisfying (2.44), a nonneqative bounded sequence $\{\delta_k\}$ and nondecreasing positive sequences $\{\varepsilon_l\}$ and $\{\eta_l\}$. Also, choose a number $\theta \in (0,1)$. Set $k := 0$, $l := 1$, $y^0 := u^0$, $j(0) := 0$, $k(0) := 0$.

Step 1 (Auxiliary procedure $\tilde{D}_k$) :

Step 1.1 : Choose q^0 from $F_{\delta_k}(u^k)$, set $i := 0$, $p^i := q^i$, $w^0 := u^k$.

Step 1.2: If

$$\|p^i\| \le \eta_l,$$

set $k(l) := k$, $y^l := u^{k+1} := u^k$, $j(k+1) := j(k)$, $k := k+1$, $l := l+1$ and go to Step 1. (*null step*)

Step 1.3: Set $w^{i+1} := u^k - \varepsilon_l p^i/\|p^i\|$, choose $q^{i+1} \in F_{\delta_k}(w^{i+1})$. If

$$\langle q^{i+1}, p^i \rangle > \theta\|p^i\|^2,$$

then set $v^k := w^{i+1}$, $g^k := q^{i+1}$, $j(k+1) := j(k)+1$, and go to Step 2. (*descent step*)

Step 1.4: Set

$$p^{i+1} := \text{Nr conv}\{p^i, q^{i+1}\},$$

$i := i+1$ and go to Step 1.2.

Step 2 (Main iteration): Set $\sigma_k := \langle g^k, u^k - v^k \rangle/\|g^k\|^2$,

$$u^{k+1} := u^k - \gamma_{j(k)}\sigma_k g^k, \tag{2.88}$$

$k := k+1$ and go to Step 1.

Thus, Method 2.3, unlike Method 2.2, need not use feasible nonexpansive operators with respect to U and its iteration sequence $\{u^k\}$ is in general infeasible. Nevertheless, it also enables one to approximate a solution of GVI (2.1). In what follows, we will call one change of the index i an inner step.

2.4.2 Convergence

In the exact case, the convergence result follows directly from Theorem 2.3.2.

Theorem 2.4.1. *Let a sequence $\{u^k\}$ be generated by Method 2.3 and let $\{\varepsilon_l\}$ and $\{\eta_l\}$ satisfy (2.54). Suppose that $\delta_k = 0$ for $k = 0, 1, \ldots$ Then:*
(i) The number of inner steps at each iteration is finite.
(ii) There exists a limit point u^* of $\{u^k\}$ which lies in S^*.
(iii) If, in addition,

$$S^* = S^*_{(d)},$$

we have

$$\lim_{k\to\infty} u^k = u^* \in S^*.$$

In accordance with Theorem 2.3.1, we can conclude that Method 2.3 also solves GVI (2.1) under Hypothesis (H2.3).

Let us now turn out to the general case. We first give some properties which can be regarded as some analogues of those in Lemmas 2.3.3 – 2.3.5. Set

$$C_k = \sup\{\|g\| \mid g \in F_{\varepsilon_l + \delta_k}(u^k)\}.$$

Lemma 2.4.2. *Let k and l be fixed. Then:*
(i) The number of inner steps is finite.
(ii) It holds that

$$p^i \in \operatorname{conv}\{q^0, \ldots, q^i\} \tag{2.89}$$

and

$$q^i, p^i \in F_{\varepsilon_l + \delta_k}(u^k) \tag{2.90}$$

for $i = 0, 1, \ldots$
(iii) It holds that

$$\begin{aligned}
\|u^{k+1} - u^*\|^2 \leq\ & \|u^k - u^*\|^2 - \gamma_{j(k)}(2 - \gamma_{j(k)})(\sigma_k \|g^k\|)^2 \\
& + 2\gamma_{j(k)}\delta_k \sigma_k C_k \quad \forall u^* \in S^*_{(d)}.
\end{aligned} \tag{2.91}$$

Proof. Assertion (i) and (2.89) follow from Lemmas 4.3.2 and 4.3.1, respectively. By construction, $q^i \in F_{\delta_k}(w^i)$ and $w^i \in B(u^k, \varepsilon_l)$, hence we have $q^i \in F_{\varepsilon_l + \delta_k}(u^k)$. Since $F_{\varepsilon_l + \delta_k}(u^k)$ is convex, we then obtain $p^i \in F_{\varepsilon_l + \delta_k}(u^k)$ because of (2.89). Thus (2.90) is true.

Next, in case $k = k(l)$, (2.91) evidently holds. Let us consider the case $k \neq k(l)$. Fix $u^* \in S^*_{(d)}$. By definition, there exist vectors $z^{i,j} \in B(w^i, \delta_k)$, $q^{i,j} \in P(z^{i,j})$ and numbers μ_j, $j \in J$, such that

$$q^i = \sum_{j \in J} \mu_j q^{i,j}, \sum_{j \in J} \mu_j = 1, \mu_j \geq 0, j \in J;$$

besides, on account of (2.87), we have $\langle q^{i,j}, z^{i,j} - u^* \rangle \geq 0$. Combining these relations gives

$$\langle q^i, w^i - u^* \rangle = \sum_{j \in J} \mu_j \left(\langle q^{i,j}, z^{i,j} - u^* \rangle + \langle q^{i,j}, w^i - z^{i,j} \rangle \right)$$

$$\geq - \sum_{j \in J} \mu_j \| q^{i,j} \| \delta_k \geq -\delta_k C_k.$$

It follows that

$$\langle g^k, u^k - u^* \rangle = \langle g^k, u^k - v^k \rangle + \langle g^k, v^k - u^* \rangle$$

$$= \sigma_k \| g^k \|^2 + \langle q^i, w^i - u^* \rangle \geq \sigma_k \| g^k \|^2 - \delta_k C_k.$$

Combining this inequality with (2.88) yields

$$\| u^{k+1} - u^* \|^2 = \| u^k - \gamma_{j(k)} \sigma_k g^k - u^* \|^2$$

$$= \| u^k - u^* \|^2 - 2\gamma_{j(k)} \sigma_k \langle g^k, u^k - u^* \rangle + (\gamma_{j(k)} \sigma_k \| g^k \|)^2$$

$$\leq \| u^k - u^* \|^2 - 2\gamma_{j(k)} \sigma_k (\sigma_k \| g^k \|^2 - \delta_k C_k) + (\gamma_{j(k)} \sigma_k \| g^k \|)^2,$$

i.e., (2.91) holds.   □

We now obtain the key properties of Method 2.3.

Proposition 2.4.2. *(i) The number of inner steps at each iteration of Method 2.3 if finite.*
(ii) Let the parameters of Method 2.3 be chosen by (2.56), (2.54) and

$$\{\delta_k\} \searrow 0. \tag{2.92}$$

If, in addition,

$$C_k \leq C' < \infty \quad \text{for } k = 0, 1, \ldots, \tag{2.93}$$

then there exist sequences $\{u^{k_s}\}$, $\{p^{k_s}\}$, and $\{\beta_{k_s}\} \searrow 0$ such that

$$\lim_{s \to \infty} p^{k_s} = 0$$

and

$$p^{k_s} \in F_{\beta_{k_s}}(u^{k_s}).$$

Proof. Part (i) follows from Lemma 2.4.2 (i). Assume for contradiction that $\{y^l\}$ is finite. Then, $l \leq t < \infty$ and there is a number $\bar{k}$ such that all the iterations of the auxiliary procedure terminate at Step 1.3 when $k \geq \bar{k}$. By construction, we then have

$$\sigma_k \| g^k \|^2 = \langle g^k, u^k - v^k \rangle = \varepsilon_l \langle q^{i+1}, p^i \rangle / \| p^i \|$$

$$\geq \theta \varepsilon_l \| p^i \| \geq \theta \varepsilon_l \eta_l \geq \tilde{\theta} > 0.$$

Applying this inequality in (2.91) and taking into account (2.56), (2.92), and (2.93), for k large enough, we obtain

$$
\begin{aligned}
\|u^{k+1} - u^*\|^2 - \|u^k - u^*\|^2 &\leq -\gamma(2-\gamma)(\sigma_k\|g^k\|)^2 + 2\gamma\delta_k\sigma_k C_k \\
&\leq -\gamma\sigma_k\left((2-\gamma)\sigma_k\|g^k\|^2 - 2\delta_k C_k\right) \\
&\leq -\gamma\sigma_k\left((2-\gamma)\tilde{\theta} - 2\delta_k C'\right) \\
&\leq -\gamma(2-\gamma)\sigma_k\tilde{\theta}/2 \\
&\leq -\gamma(2-\gamma)(\tilde{\theta}/C')^2/2,
\end{aligned}
$$

which is a contradiction. Therefore, $\{y^l\}$ is infinite. Next, at the $k(l)$th iteration, there is a number i_l such that $\|p^{i_l}\| \leq \eta_l$. Noting that $y^l = u^{k(l)}$ and using (2.90), (2.54) and (2.92), we obtain (ii) with $\beta_{k_s} = \varepsilon_l + \delta_{k(l)}$ and $u^{k_s} = y^l$.   □

It is clear that (2.93) holds if, for example,

$$
\sup\{\|g\| \mid g \in P(u), u \in R^n\} \leq C' < \infty.
$$

However, to guarantee convergence, we need additional assumptions.

Corollary 2.4.1. *Let the parameters of Method 2.3 be chosen by (2.54), (2.56) and (2.92). If $\{u^k\}$ is bounded, there exists a limit point of $\{u^k\}$ which belongs to S^*.*

Proof. Since $\{u^k\}$ is bounded, it has limit points. Moreover, as P is a K-mapping due to Proposition 2.3.1, relation (2.93) must hold. By using Proposition 2.4.2 (ii) we now conclude that $0 \in P(u^*)$, where u^* is a limit point of $\{u^k\}$.   □

In order to obtain convergence results under weaker conditions, Method 2.3 should be modified slightly. In fact, choose a convex compact set V such that $V \cap S^*_{(d)} \neq \emptyset$ and replace rule (2.88) with the following:

$$
u^{k+1} := \pi_V(u^k - \gamma_{j(k)}\sigma_k g^k). \tag{2.94}
$$

From the nonexpansive projection properties (see Proposition 1.2.1 (iii)), it follows that all the assertions of Lemma 2.4.2 and Proposition 2.4.2 with $S^*_{(d)}$ being replaced by $S^*_{(d)} \cap V$ remain valid and that $\{u^k\}$ is now bounded. Combining these assertions with Corollary 2.4.1, we obtain the following convergence result immmediately.

Theorem 2.4.2. *Let the parameters of Method 2.3 be chosen by (2.54), (2.56) and (2.92), and let rule (2.88) be substituted with (2.94). Then:*
(i) The number of inner steps at each iteration of the method is finite.
(ii) There exists a limit point of $\{u^k\}$ which lies in S^*.

In order to apply Method 2.3 under Hypothesis $(HG2.3)$, i.e., when the feasible set U is not associated to some function h it suffices to replace the mapping P in (2.48) with the mapping $\tilde{P}$ being defined in (2.84). Then, all the results of this section will be true. Indeed, we can use Proposition 2.3.2 and Theorem 2.3.5 instead of Proposition 2.3.1 and Theorem 2.3.1, respectively. Next, let us consider the problem of finding a point u^* of R^n such that

$$\forall u \in R^n, \quad \forall t \in \tilde{P}(u) \quad \langle t, u - u^* \rangle \geq 0, \tag{2.95}$$

which can be viewed as the dual problem to the multivalued inclusion

$$0 \in \tilde{P}(u^*). \tag{2.96}$$

We denote by $\tilde{S}^*_{(d)}$ the solution set of problem (2.95). Then, using Propositions 2.1.3 and 2.3.2 gives the following.

Lemma 2.4.3. *(i) $\tilde{S}^*_{(d)}$ is convex and closed.*

(ii) $\tilde{S}^*_{(d)} \subseteq \tilde{S}^*$.

(iii) If $\tilde{P}$ is pseudomonotone, then $\tilde{S}^*_{(d)} = \tilde{S}^*$.

Moreover, we can obtain an analogue of Proposition 2.4.1.

Proposition 2.4.3. $U^d = \tilde{S}^*_{(d)}$.

Proof. Take any points $u^* \in U^d$ and $u \in R^n$. If $u \notin \text{int} U$, then, by the definition of $D(u)$ in (2.82),

$$\langle t, u - u^* \rangle \geq 0 \quad \forall t \in \tilde{P}(u).$$

If $u \in U$, then for each $g \in G(u)$, we have $\langle g, u - u^* \rangle \geq 0$. On account of (2.84), we now obtain

$$\langle t, u - u^* \rangle \geq 0 \quad \forall t \in \tilde{P}(u),$$

i.e., $U^d \subseteq \tilde{S}^*_{(d)}$. The inverse inclusion is proved by the argument as in Proposition 2.4.1. □

Thus, after replacing P by $\tilde{P}$, we can apply Method 2.3 to GVI (2.1), or equivalently, to the multivalued inclusion (2.96) under Hypothesis $(HG2.3)$ as well.

2.5 Decomposable CR Method

In this section, we consider a modification of Method 2.3 for GVI (2.1) in the case where the feasible set U is a product set. Namely, denote by M the set of indexes $\{1, \ldots, m\}$. For every $u \in R^n$, let

$$u = (u_s \mid s \in M)$$

where $u_s \in R^{n_s}$ for $s \in M$, i.e. $n = \sum_{s \in M} n_s$. Accordingly, let

$$U = \prod_{s \in M} U_s, \tag{2.97}$$

where

$$U_s = \{z \in R^{n_s} \mid h_s(z) \leq 0\}, \tag{2.98}$$

$h_s : R^{n_s} \to R$ is a convex function for $s \in M$. The problem is to find an element $u^* \in U$ such that

$$\begin{aligned} \exists g_s^* \in G_s(u^*), s \in M; \\ \textstyle\sum_{s \in M} \langle g_s^*, u_s - u_s^* \rangle \geq 0 \quad \forall u_s \in U_s, s \in M; \end{aligned} \tag{2.99}$$

where $G_s : U \to \Pi(R^{n_s})$ is a K-mapping for $s \in M$. We denote by U^* the solution set of this problem. It is clear that GVI (2.99) is equivalent to GVI (2.1) with U having the form (2.97) and G having the form

$$G = (G_s \mid s \in M). \tag{2.100}$$

GVI of the form (2.99) has a number of applications, including the Nash equilibrium problem of an m-person non-cooperative game (see Section 2.1.3), traffic equilibrium and spatial price equilibrium problems (e.g., see [209, 167, 70, 154]), and, also, saddle point problems in optimization (see Corollary 1.1.2).

Throughout this section, we assume that GVI (2.99) is solvable and that,

$$\begin{aligned} \text{for every } u \in U \text{ and for every } \bar{g}_s \in G_s(u), s \in M, \\ \textstyle\sum_{s \in M} \langle \bar{g}_s, u_s - u_s^* \rangle \geq 0 \end{aligned} \tag{2.101}$$

for some $u^* \in U$. Moreover, we assume that for each $s \in M$, U_s satisfies the Slater condition, i.e., there exists a point $\bar{u}_s \in R^{n_s}$ such that $h_s(\bar{u}_s) < 0$. We denote by U^d the solution set of problem (2.101), which can be viewed as the dual problem to GVI (2.99). Thus, we have $U^d \neq \emptyset$.

Typically, GVI of the form (2.99) is quite large, hence a decomposition approach is suitable to find its solution. However, our decomposition scheme has to take into account the two-level structure of CR methods and provide team-work of both the levels.

2.5.1 Properties of Auxiliary Mappings

We first discuss condition (2.101) in detail. It is clear that (2.101) holds, if $U^* \neq \emptyset$ and G in (2.100) is monotone or pseudomonotone.

Next, for each $s \in M$, define the extended multivalued mapping $\tilde{G}_s : U \to \Pi(R^{n_s})$ by

$$\tilde{G}_s = (0, \ldots, 0, G_s, 0, \ldots, 0)^T. \tag{2.102}$$

Then we have

$$G(u) = \sum_{s \in M} \tilde{G}_s(u). \tag{2.103}$$

It is well known that the sum of pseudomonotone mappings need not be pseudomonotone. Also, if $\tilde{G}_s, s \in M$ are pseudomonotone, the same is not true for G, but (2.101) still holds for each $u^* \in U^*$ as the following example illustrates.

Example 2.5.1. Let $m = 2$, $n_1 = n_2 = 1$, $n = 2$, $G_s(u) = A_s u$ for $s = 1, 2$ where

$$A_1 = (0, 1) \text{ and } A_2 = (1, 0).$$

Also, let

$$U_s = \{u_s \in R \mid u_s \geq 0\}$$

for $s = 1, 2$. Then the mappings $\tilde{G}_s$ are quasimonotone on $U = U_1 \times U_2$ and pseudomonotone on $\mathrm{int}U$. Next, $G(u) = Au$ where

$$A = \begin{pmatrix} 0 & 1 \\ 1 & 0 \end{pmatrix},$$

and $U^d = \{(0,0)^T\} \subset U^*$. Indeed, $u_s^0 = (0) \in U_s$ and $A_s u^0 = (0,0)^T$, where $u^0 = (u_s^0 \mid s = 1, 2)$, hence $u^0 \in U^*$. Next, if we choose any $\bar{u} = (\alpha_1, \alpha_2)^T \in U$, $\bar{u} \neq u^0$, then $\alpha_i \geq 0$ for $i = 1, 2$, and

$$\langle A\bar{u}, \bar{u} - u^0 \rangle = (\alpha_2, \alpha_1)(\alpha_1, \alpha_2)^T = 2\alpha_1\alpha_2 \geq 0,$$

i.e., $\bar{u} \notin U^d$ and (2.101) holds. Take the points $u = (1,2)^T$ and $v = (2,1)^T$. Then we have $u, v \in \mathrm{int}U$ and

$$\langle Av, u - v \rangle = (1,2)(-1,1)^T = 1 > 0,$$

but

$$\langle Au, u - v \rangle = (2,1)(-1,1)^T = -1 < 0.$$

Thus, G is not quasimonotone on $\mathrm{int}U$ and it is not pseudomonotone either.

The concept of pseudomonotonicity in product sets can be modified.

Definition 2.5.1. Let the set U and the mapping G be defined by (2.97) and (2.100), respectively. The mapping G is said to be *pseudo P-monotone* on U if for each pair of points $u, v \in U$ and for all $g' \in G(u)$, $g'' \in G(v)$, we have

$$\min_{s \in M} \langle g_s'', u_s - v_s \rangle \geq 0 \quad \text{implies} \quad \langle g', u - v \rangle \geq 0.$$

Note that in the case where $m = 2$ and G is single-valued, Definition 2.1.2 reduces to [17, Definition 1]. It is clear that each pseudomonotone mapping is pseudo P-monotone, but the reverse assertion is not true in general; see [17, Example 2]. At the same time, if $u^* \in U^*$, we must have

$$\exists g_s^* \in G_s(u^*), \forall u_s \in U_s : \langle g_s^*, u_s - u_s^* \rangle \geq 0 \quad \forall s \in M.$$

Therefore, if $U^* \neq \emptyset$ and G is pseudo P-monotone, then (2.101) holds.

Proposition 2.5.1. *Let the set U be defined by (2.97). Suppose that at least one of the following assumptions is fulfilled:*
(a) The mapping G is defined by (2.102), (2.103) where $\tilde{G}_s, s \in M$ are pseudomonotone.
(b) The mapping G is defined by (2.100) and it is pseudo P-monotone. Then (2.101) holds for each $u^* \in U^*$.

Proof. Take any $u^* \in U^*$. It follows that

$$\exists g_s^* \in G_s(u^*), \forall u_s \in U_s : \langle g_s^*, u_s - u_s^* \rangle \geq 0 \quad \forall s \in M.$$

In case (b), pseudo P-monotonicity of G implies (2.101). In case (a), we have

$$\exists \tilde{g}_s \in \tilde{G}_s(u^*), \langle \tilde{g}_s, u - u^* \rangle \geq 0 \quad \forall s \in M,$$

hence, by pseudomonotonicity,

$$\forall g_s \in \tilde{G}_s(u), \forall u \in U : \langle g_s, u - u^* \rangle \geq 0 \quad \forall s \in M.$$

It follows that

$$\forall g \in G(u), \forall u \in U : \langle g, u - u^* \rangle \geq 0,$$

i.e., (2.101) also holds. $\square$

Define the mapping F_s from $U \times R^{n_s}$ into R^{n_s} by

$$F_s(u, v_s) = \begin{cases} G_s(u\|v_s) & \text{if } h_s(v_s) < 0, \\ \text{conv}\{G_s(u, v_s) \cup \partial h_s(v_s)\} & \text{if } h_s(v_s) = 0, \\ \partial h_s(v_s) & \text{if } h_s(v_s) > 0; \end{cases} \quad (2.104)$$

where

$$(u\|v_s) = (u_1, \ldots, u_{s-1}, v_s, u_{s+1}, \ldots, u_m).$$

It is clear that the mappings F_s are analogues of P in (2.48) for the case of (2.97), (2.98). Therefore, the analogues of Proposition 2.3.1 and Theorem 2.3.1 are true for F_s in (2.104).

Proposition 2.5.2. *For each $s \in M$, F_s is a K-mapping.*

Proof. By definition, F_s has nonempty convex and compact values. Next, G_s and ∂h_s are u.s.c. Due to the properties of u.s.c. mappings (e.g., see [210, Theorem 5.1.4]), F_s is also u.s.c. □

Proposition 2.5.3. *A point $u^* = (u_s^* \mid s \in M)$ belongs to U^* if and only if*

$$0 \in F_s(u^*, u_s^*) \qquad \forall s \in M. \tag{2.105}$$

Proof. Following the proof of Theorem 2.3.1 we see that (2.105) is equivalent to

$$u_s^* \in U_s, \quad 0 \in \begin{cases} G_s(u^*) & \text{if } h_s(u_s^*) < 0, \\ G_s(u^*) + \text{cone}\partial h_s(u_s^*)\} & \text{if } h_s(u_s^*) = 0. \end{cases}$$

for all $s \in M$. From Lemma 2.3.2 it now follows that (2.105) is equivalent to

$$u_s^* \in U_s, 0 \in G_s(u^*) + N(U_s, u_s^*) \quad \forall s \in M.$$

□

2.5.2 Description of the Method

In order to construct a CR method with decomposition techniques we cannot apply the usual decomposition schemes (e.g., see [167]) directly because of the two-level structure of the method and of the binding condition (2.46) (see also Lemma 2.3.4 (iii)). To overcome these difficulties, we propose to apply different and independent decomposition schemes for both levels. We now describe the main part of the modified auxiliary procedure, which can be viewed as some modification of that in Step 1 of Method 2.2.

Fix integers $k \geq 0$, $l \geq 0$, $s \in M$ and a point $u \in U$. Define the mapping $Q_s : U \times R^{n_s} \to \Pi(R^{n_s})$ by

$$Q_s(u, v_s) = \begin{cases} G_s(u\|v_s) & \text{if } h_s(v_s) \leq 0, \\ \partial h_s(v_s) & \text{if } h_s(v_s) > 0. \end{cases}$$

Procedure $D_{k,s}^l$. *Data:* A point $u^k \in U$.
Output: Vectors g_s^k, v_s^k, a number $\bar{\sigma}_{k,s}$.
Parameters: Numbers $\theta \in (0, 1)$, $\varepsilon_l > 0$, $\eta_l > 0$.
Step 1: Choose q^0 from $Q_s(u^k, u_s^k)$, set $i := 0$, $p^i := q^i$, $w^i := u_s^k$.
Step 2: If $\|p^i\| \geq \eta_l$, go to Step 3. Otherwise, set $g_s^k := 0$, $v_s^k := u_s^k$, $\bar{\sigma}_{k,s} := 0$ and stop.
Step 3: Set $w^{i+1} := w^0 - \varepsilon_l p^i / \|p^i\|$, choose q^{i+1} from $Q_s(u^k, w^{i+1})$. If

$$\langle q^{i+1}, p^i \rangle > \theta \|p^i\|^2,$$

then set $g_s^k := q^{i+1}$, $v_s^k := w^{i+1}$, $\bar{\sigma}_{k,s} := \langle g_s^k, u_s^k - v_s^k \rangle$ and stop.
Step 4: Set $p^{i+1} := \text{Nr conv}\{p^i, q^{i+1}\}$, $i := i + 1$ and go to Step 2.

We will call one change of the index i an inner step. From Lemmas 4.3.1 and 4.3.2 we now have the following.

Lemma 2.5.1. *(i) The number of inner steps in Procedure $D_{k,s}^l$ is finite.*
  (ii)

$$p^i \in \text{conv} \bigcup_{z_s \in B(u_s^k, \varepsilon_l)} F_s(u^k, z_s), \qquad (2.106)$$

for $i = 0, 1, \ldots$
  (iii)

$$\|q^i\| \le C_1 < \infty$$

for $i = 0, 1, \ldots$

The CR method for solving GVI (2.99) can be described as follows.

Method 2.4. *Step 0 (Initialization)*: Choose a point $u^0 \in U$, a sequence $\{\gamma_j\}$ satisfying (2.44) and sequences $\{\varepsilon_l\}$ and $\{\eta_l\}$ satisfying (2.54).
  Also, choose a number $\theta \in (0, 1)$, and, for each $s \in M$, choose a sequence of mappings $\{P_{k,s} : R^{n_s} \to U_s\}$ such that $P_{k,s} \in \mathcal{F}(U_s)$ for $k = 0, 1, \ldots$ Set $k := 0$, $l := 1$, $y^0 := u^0$, $j(0) := 0$, $k(0) := 0$.
  Step 1 (Auxiliary procedure $\tilde{D}_k$) :
    Step 1.1 : Apply Procedure $D_{k,s}^l$. If $\bar{\sigma}_{k,s} > 0$, go to Step 1.3.
    Step 1.2: If $s < m$, set $s := s + 1$ and go to Step 1. Otherwise, set $u^{k+1} := v^k := u^k$, $\sigma_k := \bar{\sigma}_{k,m}$, $g^k := 0$, $k(l) := k$, $y^l := u^k$, $l := l + 1$, $j(k + 1) := j(k)$, and go to Step 1. *(null step)*
    Step 1.3: Set $v^k := (u^k \| v_s^k)$, compute

$$g_i^k \in G_i(v^k) \quad \text{for } i \in M, i \ne s.$$

Set $j(k + 1) := j(k) + 1$, $\sigma_k := \bar{\sigma}_{k,s} / \|g^k\|^2$. *(descent step)*
  Step 2 (Main iteration): Set

$$u_i^{k+1} := P_{k,i}(u_i^k - \gamma_{j(k)} \sigma_k g_i^k, U_i), \quad i \in M,$$

$k := k + 1$ and go to Step 1.

By construction, Method 2.4 is different from Method 2.2 in case $m > 1$. In particular, in order to provide for the key property (i) of Lemma 2.3.5 to hold, Method 2.2 involves an auxiliary procedure in the space R^n, whereas Method 2.4 replaces such a procedure with a sequence of Procedures $D_{k,s}^l$, which are carried out in the corresponding subspaces R^{n_s}. Nevertheless, even after such a reduced step, one has a hyperplane separating the current iterate and the solution set, i.e., Method 2.4 will also maintain the monotone convergence to a solution. Next, Method 2.4 corresponds to a consequentive decomposition scheme, Steps 1 and 2 resembling the coordinate descent processes. However, after a slight modification, we can apply the parallel decomposition scheme for Steps 1 and 2 as well. Namely, all Procedures $D_{k,s}^l$ can be carried out in parallel until either a descent or a null step occurs.

2.5.3 Convergence

The following lemma is an analogue of Lemma 2.3.5 for Method 2.4.

Lemma 2.5.2. *(i) For every $u^* \in U^d$, we have*

$$\|u^{k+1} - u^*\|^2 \leq \|u^k - u^*\|^2 - \gamma_{j(k)}(2 - \gamma_{j(k)})(\sigma_k\|g^k\|)^2 \qquad (2.107)$$

for $k = 0, 1, \ldots$
  (ii) Sequences $\{u^k\}$, $\{v^k\}$ and $\{g^k\}$ are bounded.
  (iii) If $l \leq \bar{l} < \infty$ for $k = 0, 1, \ldots$, then

$$\liminf_{k \to \infty} (\sigma_k\|g^k\|)^2 = 0.$$

  (iv) If $\sigma_k\|g^k\| \geq \tilde{\sigma} > 0$ as $l \leq \bar{l} < \infty$, the sequence $\{y^l\}$ is infinite.
  (v) For each limit point u^* of $\{u^k\}$ such that $u^* \in U^d$ we have

$$\lim_{k \to \infty} u^k = u^*.$$

Proof. Let us consider the kth iteration, $k = 0, 1, \ldots$ It is clear that relation (2.107) holds in the case of $\sigma_k = 0$. Otherwise, let $\sigma_k > 0$. Then there exists $s \in M$ such that $\bar{\sigma}_{k,s} > 0$. By using the argument as in Lemma 2.3.4 (iii), we conclude that $v_s^k \in U_s$, hence $v^k \in U$ and $g_s^k \in G_s(v^k)$. For every $u^* \in U^d$, by definition, we then obtain

$$
\begin{aligned}
\langle g^k, u^k - u^* \rangle &= \langle g^k, u^k - v^k \rangle + \langle g^k, v^k - u^* \rangle \\
&\geq \langle g_s^k, u_s^k - v_s^k \rangle = \bar{\sigma}_{k,s} = \sigma_k\|g^k\|^2.
\end{aligned}
$$

Using the argument as in Lemma 1.2.3, we now obtain (2.107) as well. By (2.107), $\{u^k\}$ is bounded, and, by construction, so are $\{v^k\}$ and $\{g^k\}$. Part (iii) follows from (2.107) and (2.44). Parts (iv) and (v) follow from parts (iii) and (i), respectively. $\square$

We now establish a convergence result for Method 2.4.

Theorem 2.5.1. *Let a sequence $\{u^k\}$ be generated by Method 2.4. Then:*
  (i) The number of inner steps at each iteration is finite.
  (ii) There exists a limit point u^* of $\{u^k\}$ which lies in U^*.
  (iii) If, in addition, $U^* = U^d$, we have

$$\lim_{k \to \infty} u^k = u^* \in U^*.$$

Proof. Part (i) follows from Lemmas 2.5.1 (i) and 2.5.2 (ii) in view of the construction of Method 2.4. Next, if $l \leq t < \infty$, then there exists a number $\bar{k}$ such that all the iterations of the auxiliary procedure terminate at Step 1.3 when $k \geq \bar{k}$. By definition and by Lemma 2.5.2 (ii), we now have

$$\sigma_k\|g^k\| = \langle g_s^k, u_s^k - v_s^k\rangle/\|g^k\| \geq \theta\varepsilon_t\eta_t/C_2 = \tilde{\sigma} > 0,$$

and we see that $\{y^l\}$ is infinite due to Lemma 2.5.2 (iv). Moreover, by Lemma 2.5.2 (ii), $\{y^l\}$ is bounded, hence, there exists a subsequence of $\{y^l\}$ that converges to some point $u^* \in U$. Without loss of generality we suppose that

$$\lim_{l\to\infty} y^l = u^*.$$

At the $k(l)$th iteration, for each $s \in M$, there exists a number $i = i(s)$ such that $\|p^i\| \leq \eta_l$. Using (2.54), (2.106) with $k = k(l)$ and Proposition 2.5.2 now yields (2.105), i.e., $u^* \in U^*$ due to Proposition 2.5.3. This proves part (ii). Part (iii) now follows from part (ii) and Lemma 2.5.2 (v). The proof is complete.   □

By using the results of Section 2.3.5, we can adjust Method 2.4 to solve GVI (2.99) in the case where the sets U_s are not associated to some functions. It suffices to make the corresponding substitutions of the mappings Q_s and F_s.

3. Applications and Numerical Experiments

In this chapter, we discuss possible applications of CR methods, especially to economic equilibrium problems. There exist several excellent books and surveys which are devoted to various applications of VI's; e.g., see [9, 45, 55, 66, 70, 80, 154, 166, 212]. But, on the one hand, the description of all possible applications of CR methods would force us to include great supplementary material and to essentially increase the volume of this book. We hope that the results of the previous chapters and the descriptions from the works above enable one to evaluate most of such applications. On the other hand, we have proved that CR methods are convergent under rather mild assumptions and we would like to reveal their properties for a broad class of applied problems. For this reason, we restrict ourself with economic applications and, besides, choose typical test problems, which model usual properties of problems in other areas. Of course, we compare results for CR methods with those of other iterative methods having similar convergence properties. To make our work more self-contained we give the description of most such methods in the first section of this chapter.

3.1 Iterative Methods for Non Strictly Monotone Variational Inequalities

It was noticed in Section 1.2, that most methods for VI's constructed by analogy with well-known optimization methods are convergent under assumptions which are stronger than the monotonicity of the underlying mapping. On the other hand, CR methods need not such restrictive assumptions for convergence. In this section, we briefly describe other iterative solution methods for (generalized) VI's which are convergent provided the underlying mapping is monotone. These methods were investigated rather well so that we give only the results which make their convergence properties more clear.

3.1.1 The Proximal Point Method

We first describe methods for variational inequalities with multivalued mappings. Namely, the blanket assumptions are the following.

Hypothesis (H3.1)

(a) U *is a nonempty, closed and convex subset of* R^n;
(b) $G : U \to \Pi(R^n)$ *is a monotone K-mapping;*
(c) GVI (2.1) *is solvable, i.e.,* $U^* \neq \emptyset$.

Then the proximal point method can be described as follows.

Method 3.1 (PPM). Choose a point $u^0 \in U$, a number $\theta > 0$. Given u^k, the next iterate u^{k+1} is defined as a solution of the following problem:

$$\exists g^{k+1} \in G(u^{k+1}), \ \langle g^{k+1} + \theta^{-1}(u^{k+1} - u^k), w - u^{k+1} \rangle \geq 0 \quad \forall w \in U. \quad (3.1)$$

Thus, PPM consists of replacing the initial problem (2.1) with a sequence of regularized problems (3.1). Note that the cost mapping in (3.1) is strongly monotone, hence, by Proposition 2.1.5 (ii), problem (3.1) always has a unique solution. Generally, for each $u \in U$, one can define the point $v(u)$ as a solution of the problem:

$$\exists g \in G(v(u)), \ \langle g + \theta^{-1}(v(u) - u), w - v(u) \rangle \geq 0 \quad \forall w \in U, \quad (3.2)$$

thus obtaining the single-valued proximal mapping $u \mapsto v(u)$. Convergence of PPM is based on the following properties of the proximal mapping.

Lemma 3.1.1. *(i)* $u^* \in U^* \iff u^* = v(u^*)$.
(ii) For all $u', u'' \in U$, *we have*

$$\langle v(u'') - v(u'), u'' - u' \rangle \geq \|v(u'') - v(u')\|^2. \quad (3.3)$$

(iii) The mapping $u \mapsto v(u)$ *is continuous.*

Proof. (i) If $u^* = v(u^*) \in U$, then (3.2) implies (2.1), i.e., $u^* \in U^*$. Conversely, let $u^* \in U^*$, but $u^* \neq v(u^*)$. Applying (3.2) with $w = u^*$ then gives

$$\langle g, u^* - v(u^*) \rangle \geq \theta^{-1} \|v(u^*) - u^*\|^2 > 0.$$

Since G is monotone, it follows that

$$\langle g^*, u^* - v(u^*) \rangle \geq \langle g, u^* - v(u^*) \rangle > 0$$

for every $g^* \in G(u^*)$, i.e., $u^* \notin U^*$, a contradiction.
(ii) Adding (3.2) with $u = u'$, $w = u''$ and (3.2) with $u = u''$, $w = u'$ gives

$$\langle g' \ - \ g'', v(u'') - v(u') \rangle$$
$$+ \ \theta^{-1} \langle [v(u') - v(u'')] - (u' - u''), v(u'') - v(u') \rangle \geq 0.$$

for some $g' \in G(v(u'))$ and $g'' \in G(v(u''))$. Since G is monotone, the above inequality yields (3.3).

(iii) Applying (3.3) and the Cauchy-Schwarz inequality gives

$$\|u'' - u'\| \geq \|v(u'') - v(u')\|,$$

i.e., v is continuous, as desired.   □

For each $u \in U$, set

$$F(u) = \theta^{-1}(u - v(u)).$$

By Lemma 3.1.1, the mapping F is continuous, and

$$u^* \in U^* \Longleftrightarrow F(u^*) = 0.$$

Moreover, F possesses an additional strong monotonicity type property, as the following lemma states.

Lemma 3.1.2. *For all $u', u'' \in U$, we have*

$$\langle F(u'') - F(u'), u'' - u' \rangle \geq \theta \|F(u'') - F(u')\|^2. \tag{3.4}$$

Proof. Fix $u', u'' \in U$. Then, using (3.3) gives

$$
\begin{aligned}
\langle F(u'') - F(u'), u'' - u' \rangle &= \theta \langle F(u'') - F(u'), F(u'') - F(u') \rangle \\
&+ \langle F(u'') - F(u'), v(u'') - v(u') \rangle \\
&= \theta \|F(u'') - F(u')\|^2 \\
&+ \theta^{-1}(\langle u'' - u', v(u'') - v(u') \rangle \\
&- \|v(u'') - v(u')\|^2) \\
&\geq \theta \|F(u'') - F(u')\|^2,
\end{aligned}
$$

and the result follows.   □

Due to the property (3.4), the usual method

$$u^{k+1} = u^k - \theta F(u^k) = v(u^k)$$

has to converge to a point u^* such that $F(u^*) = 0$, i.e., to a solution of GVI (2.1). However, the above method is evidently coincides with PPM. We give the corresponding convergence result in the following theorem.

Theorem 3.1.1. *If a sequence $\{u^k\}$ is generated by PPM, then*

$$\lim_{k \to \infty} u^k = u^* \in U^*. \tag{3.5}$$

Proof. Fix $v^* \in U^*$. Then $F(v^*) = 0$. Applying (3.4) now gives

$$
\begin{aligned}
\|u^{k+1} - v^*\|^2 &= \|u^k - \theta F(u^k) - (v^* - \theta F(v^*))\|^2 \\
&= \|u^k - v^*\|^2 - 2\theta \langle F(u^k) - F(v^*), u^k - v^* \rangle \\
&+ \theta^2 \|F(u^k) - F(v^*)\|^2 \\
&\leq \|u^k - v^*\|^2 - \theta^2 \|F(u^k) - F(v^*)\|^2.
\end{aligned}
$$

It follows that $\{u^k\}$ is bounded and that

$$\lim_{k \to \infty} \|u^k - v(u^k)\| = 0.$$

Hence, $\{u^k\}$ has limit points, and, without loss of generality, we suppose $\lim_{k \to \infty} u^k = u^*$. Combining the above relations and Lemma 3.1.1 (iii), we obtain $u^* = v(u^*) \in U^*$ due to Lemma 3.1.1 (i). Since $\|u^k - u^*\|$ is monotone decreasing, we get (3.5). □

Using the proof of Theorem 3.1.1 we can conclude that the convergence rates of PPM are similar to those of the methods from Sections 1.3 and 1.4. For example, PPM converges at least linearly if G is strongly monotone. The main drawback of PPM consists in the great computational expenses per iteration. In fact, the auxiliary problem (3.1) merely slightly different from the initial problem (2.1). Therefore, the auxiliary problem (3.1) is usually solved approximately. Besides, there are simplified versions of PPM for several specific classes of (G)VI's.

3.1.2 Regularization and Averaging Methods

We now describe simple methods for solving GVI (2.1) under Hypothesis (H3.1). These methods modify the usual projection method:

$$u^{k+1} = \pi_U(u^k - \alpha_k g^k), \quad g^k \in G(u^k), \tag{3.6}$$

which fails to provide convergence under Hypothesis (H3.1); see Example 1.2.1. The idea of the first method consists in applying the method (3.6) to the problem (2.1) after adding a strongly monotone mapping to G. It is called an iterative regularization method(IRM).

Method 3.2 (IRM). Choose a point $u^0 \in U$ and positive sequences $\{\alpha_k\}$ and $\{\theta_k\}$. Given u^k, the next iterate u^{k+1} is defined as follows:

$$u^{k+1} := \pi_U(u^k - \alpha_k(g^k + \theta_k u^k)), \quad g^k \in G(u^k). \tag{3.7}$$

It is clear that (3.7) is close to (3.6) if θ_k is sufficiently small. On the other hand, if $\theta_k > 0$ is fixed, when, under a suitable choice of $\{\alpha_k\}$, the sequence $\{u^k\}$ converges to a unique solution of the perturbed problem:

$$\exists g \in G(u_\theta), \quad \langle g + \theta u_\theta, u - u_\theta \rangle \geq 0 \quad \forall u \in U.$$

Thus, if $\{\theta_k\} \to 0$ as $k \to \infty$, then we can expect that the method (3.7) will converge to a solution of the initial problem, as the following proposition states.

Proposition 3.1.1. *[11, Theorem 3.1] Suppose that there exists a constant L such that*

$$\|g\| \le L(1 + \|u\|) \quad \forall g \in G(u), \quad \forall u \in U.$$

Also, suppose that $\{\alpha_k\}$ and $\{\theta_k\}$ satisfy the conditions:

$$\lim_{k\to\infty} \theta_k = 0, \; \lim_{k\to\infty} \alpha_k/\theta_k = 0,$$
$$\lim_{k\to\infty} (\theta_k - \theta_{k-1})/(\alpha_k \theta_k^2) = 0, \; \sum_{k=0}^{\infty} \theta_k \alpha_k = \infty. \tag{3.8}$$

Then IRM generates a sequence $\{u^k\}$ such that

$$\lim_{k\to\infty} \|u^k - \mathrm{Nr}U^*\| = 0.$$

Note that assumptions (3.8) are fulfilled if, for example,

$$\alpha_k = (k+1)^{-1/2}, \theta_k = (k+1)^{-\tau}, \tau \in (0, 1/2).$$

Nevertheless, the convergence of IRM can be rather slow. More precisely, it was shown in [11, Chapter 3], that $\|u^k - \mathrm{Nr}U^*\| = O(\theta_k)$.

Another modification of method (3.6) consists in replacing the usual convergence of $\{u^k\}$ with an ergodic convergence. In other words, we consider the sequence

$$z^k = \sum_{i=0}^{k} \alpha_i u^i / \sum_{i=0}^{k} \alpha_i, \tag{3.9}$$

which enjoys stronger convergence properties than $\{u^k\}$. This idea leads to the so-called averaging method, which can be described as follows.

Method 3.3 (AM). Choose a point $u^0 \in U$ and a positive sequence $\{\alpha_k\}$. Set $z^0 := u^0$, $\beta_0 := \alpha_0$. At the kth iteration, $k = 0, 1, \ldots$, set

$$\begin{aligned}
\beta_{k+1} &:= \beta_k + \alpha_{k+1}, \tau_{k+1} = \alpha_{k+1}/\beta_{k+1}; \\
u^{k+1} &:= \pi_U(u^k + \alpha_k g^k), g^k \in G(u^k); \\
z^{k+1} &:= \tau_{k+1} u^{k+1} + (1 - \tau_{k+1}) z^k.
\end{aligned}$$

From the description it follows that the sequence $\{z^k\}$ generated by AM satisfies (3.9). Since convergence properties of AM are not obvious, we give the full proof of the corresponding convergence result.

Proposition 3.1.2. *Suppose that sequences $\{u^k\}$ and $\{z^k\}$ are constructed by AM and that the sequence $\{\alpha_k\}$ satisfies the following conditions:*

$$\sum_{k=0}^{\infty} \alpha_k = \infty, \quad \sum_{k=0}^{\infty} (\alpha_k \|g^k\|)^2 < \infty. \tag{3.10}$$

Then there exists limit points of $\{z^k\}$ and all these points lie in U^*.

Proof. Take any $u \in U$. By construction, we have

$$\|u^{i+1} - u\|^2 \le \|u^i - \alpha_i g^i - u\|^2 = \|u^i - u\|^2 - 2\alpha_i \langle g^i, u^i - u \rangle + \alpha_i^2 \|g^i\|^2.$$

Using the monotonicity of G gives

$$\|u^{i+1} - u\|^2 \le \|u^i - u\|^2 - 2\alpha_i \langle g, u^i - u \rangle + \alpha_i^2 \|g^i\|^2 \qquad (3.11)$$

for every $g \in G(u)$. Applying this inequality with $u = u^* \in U^*$ gives

$$\|u^{i+1} - u^*\|^2 \le \|u^i - u^*\|^2 + \alpha_i^2 \|g^i\|^2$$

due to (2.1). Taking into account (3.10) we conclude that $\{u^k\}$ is bounded, hence, so is $\{z^k\}$. Therefore, $\{z^k\}$ has limit points. Let z^* be a limit point of $\{z^k\}$, i.e., $\lim_{l \to \infty} z^{k_l} = z^*$. Summing (3.11) over $i = 0, 1, \ldots, k$ gives

$$\|u^{i+1} - u\|^2 \le \|u^0 - u\|^2 - 2\langle g, \sum_{i=0}^{k} \alpha_i(u^i - u) \rangle + \sum_{i=0}^{k} \alpha_i^2 \|g^i\|^2,$$

or equivalently,

$$\langle g, z^k - u \rangle \le \left(\|u^0 - u\|^2 - \|u^{k+1} - u\|^2 + \sum_{i=0}^{k} \alpha_i^2 \|g^i\|^2 \right) \bigg/ \left(2 \sum_{i=0}^{k} \alpha_i \right).$$

Taking the limit $k = k_l \to \infty$ in this inequality gives

$$\langle g, z^* - u \rangle \le 0 \quad \forall g \in G(u), \forall u \in U.$$

Hence, $z^* \in U^d$. On account of Proposition 2.1.3 (ii), we have $z^* \in U^*$. The proof is complete. $\qquad \square$

Thus, the averaging method converges under the monotonicity assumption on G. Its rate of convergence was investigated by several authors; e.g., see [155, 212]. Note that it can be rather slow because of the first relation in (3.10). More precisely, it was shown in [155] that $\|z^k - u^*\| = O(1/\sqrt{k})$ where $u^* \in U^*$, i.e., the rate of convergence is similar to that of IRM.

It should be noted that the monotonicity assumption is essential for convergence of Methods 3.1 – 3.3, i.e., the question whether they are convergent under weaker assumptions or not is still open.

3.1.3 The Ellipsoid Method

We consider GVI (2.1) under Hypothesis $(H2.3)$ of Section 2.3, i.e., the cost mapping G now need not be monotone. The idea of the ellipsoid method consists in constructing a sequence of ellipsoids $\{E_k\}$, which contain a point of U^d, with $\mathrm{vol}(E_k)$ tending to zero as $k \to \infty$. The implementable version of the ellipsoid method is based on the following observations.

Let z be a point in R^n and let D be an $n \times n$ positive definite matrix. Then one can define the ellipsoid ell(D, z) as follows:

$$\text{ell}(D, z) = \{x \in R^n \mid \langle D^{-1}(x - z), x - z \rangle \leq 1\}.$$

It is clear that ell$(I, 0) = B(0, 1)$.

Proposition 3.1.3. *[197, Theorem 13.1] The ellipsoid* ell(D', z'), *where*

$$z' = z - \frac{1}{n+1} \frac{Da}{\sqrt{\langle a, Da \rangle}},$$

$$D' = \frac{n^2}{n^2 - 1} \left(D - \frac{2}{n+1} \frac{Daa^T D^T}{\langle a, Da \rangle} \right), \tag{3.12}$$

has the minimal volume among all those containing the half-ellipsoid

$$\text{ell}(D, z) \bigcap \{x \in R^n \mid \langle a, x - z \rangle \leq 0\}. \tag{3.13}$$

Moreover,

$$\frac{\text{vol}[\text{ell}(D', z')]}{\text{vol}[\text{ell}(D, z)]} = \left(\frac{n^2}{n^2 - 1} \right)^{\frac{n-1}{2}} \times \left(\frac{n}{n+1} \right) < e^{-\frac{1}{2(n+1)}}. \tag{3.14}$$

Thus, the volume of the ellipsoid ell(D', z') will reduce if we make use of (3.12) to find the next ellipsoid. It suffices to find the rule of choosing the element a.

Lemma 3.1.3. *Suppose that* $u^* \in$ ell$(D, z) \bigcap U^d$. *If we set* a *in (3.13) to be an element of* $Q(z)$ *being defined in (2.43), then* $u^* \in$ ell(D', z').

Proof. From the definition of Q it follows that for each $q \in Q(z)$,

$$u^* \in \{x \in R^n \mid \langle q, x - z \rangle \leq 0\}.$$

Hence, the half-ellipsoid (3.13) with $a = q$ contains u^* and, evidently, so is ell(D', z'). □

Taking Lemma 3.1.3 and Proposition 3.1.3 as a basis, we can construct an iteration sequence as follows.

Method 3.4 (EM). Choose a point $u^0 \in R^n$, a number $\theta > 0$ such that $U^d \bigcap B(u^0, \theta) \neq \emptyset$ and set $D_0 := \theta^2 I$. At the kth iteration, $k = 0, 1, \ldots$, choose $q^k \in Q(u^k)$, set

$$u^{k+1} := u^k - \frac{1}{n+1} \frac{D_k q^k}{\sqrt{\langle q^k, D_k q^k \rangle}},$$

$$D_{k+1} := \frac{n^2}{n^2-1} \left(D_k - \frac{2}{n+1} \frac{D_k q^k (D_k q^k)^T}{\langle q^k, D_k q^k \rangle} \right),$$ (3.15)

and $k := k+1$.

To justify EM it now suffices to observe that the recurrence (3.15) maintains the positive definiteness of D_{k+1}. Therefore, we must have

$$U^d \bigcap \text{ell}(D_k, u^k) \neq \emptyset$$

and $\text{vol}(\text{ell}(D_k, u^k)) \to 0$ as $k \to \infty$. These properties yield the convergence of EM. Moreover, the estimate (3.14) allows one to derive a rate of convergence. For each $u \in R^n$, set

$$\varphi(u) = \sup\{\langle q, u - v \rangle \mid q \in Q(v), v \in R^n\}.$$ (3.16)

Using Proposition 2.4.1, we conclude that $\varphi(u) \geq 0$ for each $u \in R^n$ and that $\varphi(u^*) = 0$ implies $u^* \in U^d \subseteq U^*$. Therefore, φ can be viewed as a gap function for GVI (2.1).

Proposition 3.1.4. *[155, Theorem 1] The total number of iterations which guarantees for EM to obtain a point u such that $\varphi(u) \leq \varepsilon$ does not exceed the value*

$$Cn^2 \ln(1/\varepsilon),$$

where C is a constant which is independent of n and ε.

Thus, the ellipsoid method converges to a solution in a linear rate with respect to the merit function (3.16), the ratio depending on the dimension n rather than the constant of strong monotonicity.

3.1.4 The Extragradient Method

We now consider a method for VI (1.1), i.e., for the single-valued case. More precisely, we suppose that Hypothesis $(H1.3)$ of Section 1.3 holds and also, $G : U \to R^n$ is a Lipschitz continuous mapping, $U^* = U^d \neq \emptyset$. The idea of the extragradient method consists in replacing the usual projection iteration (3.6) with the double projection iteration. More precisely, it can be described as follows.

Method 3.5 (EGM). Choose a point $u^0 \in U$, a number $\beta > 0$. Given u^k, the next iterate u^{k+1} is defined as follows:

$$u^{k+1} := \pi_U(u^k - \beta G(v^k)), \quad v^k := \pi_U(u^k - \beta G(u^k)).$$ (3.17)

Note that if G is not strictly monotone, then we have

$$\langle G(u), u - u^* \rangle \geq 0 \quad \forall u^* \in U^*,$$

but the strict inequality, generally speaking, does not hold. In other words, the angle between $-G(u^k)$ and $u^* - u^k$ need not be acute, so that the usual projection method

$$u^{k+1} := \pi_U(u^k - \beta G(g^k))$$

fails to provide convergence. On the other hand, the second relation in (3.17) can be rewritten as follows:

$$\langle G(u^k) + \beta^{-1}(v^k - u^k), v - v^k \rangle \geq 0 \quad \forall v \in U.$$

Setting $v = u^k$ in this inequality gives

$$\langle G(u^k), u^k - v^k \rangle \geq \beta^{-1} \|u^k - v^k\|^2.$$

Therefore, for $u^* \in U^*$, we have

$$
\begin{aligned}
\langle G(v^k), u^k - u^* \rangle &= \langle G(v^k), u^k - v^k \rangle + \langle G(v^k), v^k - u^* \rangle \\
&\geq \langle G(u^k), u^k - v^k \rangle + \langle G(v^k) - G(u^k), u^k - v^k \rangle \\
&\geq \beta^{-1}\|u^k - v^k\|^2 - L\|u^k - v^k\|^2 \geq \beta'\|u^k - v^k\|^2 > 0,
\end{aligned}
$$

if $\beta < 1/L$, where L is the Lipschitz constant for G. Thus $-G(v^k)$ makes an acute angle with $u^* - u^k$, so that the process (3.17), under a suitable choice of β, will converge to a solution of VI (1.1).

Although EGM and CR methods are in general based on different ideas, it is easy to see that EGM is close to the simplest variant of Methods 1.1 and 1.2, where (1.27) holds with $A_k \equiv I$. Nevertheless, EGM, unlike Methods 1.1 and 1.2, requires the Lipschitz constant L to be evaluated for convergence (see [128, 3]). In [88], a variant of the extragradient method with linesearch was proposed. It corresponds (3.17) with $\beta = \tilde{\beta}\theta^m$ where $\tilde{\beta} > 0$ and m is the smallest number in Z_+ such that

$$\tilde{\beta}\theta^m \leq \frac{\alpha\|u^k - \pi_U(u^k - \tilde{\beta}\theta^m G(u^k))\|}{\|G(u^k) - G(\pi_U(u^k - \tilde{\beta}\theta^m G(u^k)))\|}, \quad \alpha \in (0,1). \tag{3.18}$$

It is clear that rule (3.18) is different from the lineasearch in CR methods. In particular, it computes the same stepsize for both steps in (3.17). The rates of convergence of EGM are similar to those of CR methods.

Proposition 3.1.5. [211, 3] *Let a sequence $\{u^k\}$ be generated by process (3.17) with $\beta < 1/L$.*

(i) If Assumption (A5) holds with $T_k(u, z) = z - u$, then $\{u^k\}$ converges to a point of U^* in a linear rate.

(ii) If Assumption (A6) holds, then process (3.17) terminates with a solution.

In [180], a variant of EGM for finding saddle points of a nondifferentiable convex-concave function was proposed. Its stepsize rule is close to (3.10), hence its convergence rates are close to that of the averaging method.

3.2 Economic Equilibrium Problems

In this section, we consider applications of CR methods to several well-known economic equilibrium models and give some results of numerical experiments illustrating these applications.

3.2.1 Economic Equilibrium Models as Complementarity Problems

A number of economic models are adjusted to investigating the conditions which balance supply and demand of commodities, i.e., they are essentially equilibrium models. As a rule, the concept of economic equilibrium can be written in terms of a complementarity relation between the price and the excess demand for each commodity. Therefore, most existing economic equilibrium models can be written as complementarity (or variational inequality) problems. To illustrate this assertion, we now describe one of the most general economic models originated by L. Walras [216].

So, it is assumed that our economy deals in n commodities and that there are m economics agents dealing with these commodities. Let $M = \{1, \ldots, m\}$. We divide M into two subsets M_s and M_c which correspond to sectors (producers) and consumers, respectively. Given a price vector $p \in R^n_+$, the jth sector determines its supply set $S_j(p) \subseteq R^n_+$ and the ith consumer determines its demand set $D_i(p) \subseteq R^n_+$. Set

$$S(p) = \sum_{j \in M_s} S_j(p), D(p) = \sum_{i \in M_c} D_i(p). \tag{3.19}$$

Then we can define the excess demand mapping

$$E(p) = D(p) - S(p). \tag{3.20}$$

A vector p^* is said to be an *equilibrium price* if it satisfies the following conditions:

$$p^* \in R^n_+; \quad \exists q^* \in E(p^*): \quad -q^* \in R^n_+, \quad \langle q^*, p^* \rangle = 0. \tag{3.21}$$

We denote by P^* the whole set of equilibrium prices. If we now set

$$G(p) = -E(p), \tag{3.22}$$

then problem (3.21) obviously coincides with GCP (2.3) or equivalently, with the following GVI: find $p^* \in R^n_+$ such that

$$\exists g^* \in G(p^*), \quad \langle g^*, p - p^* \rangle \geq 0 \quad \forall p \in R^n_+; \tag{3.23}$$

see Proposition 2.1.6. Therefore, one can derive the existence and uniqueness conditions for problem (3.23) (or (3.21)) directly from Propositions 2.1.2 and

2.1.5. Moreover, a number of additional existence and uniqueness results, essentially exploiting features of economic equilibrium models, have been obtained; e.g., see [6, 22, 159, 210]. On the other hand, there are a few approaches to compute economic equilibria. Most works are devoted to Scarf's simplicial labeling method [193], its modifications and extensions. Such methods are applicable for general economic equilibrium problems, however, need considerable additional storage, hence, they are implementable for rather small problems. Another approach is based on Newton's type methods; see [144, 148]. However, to ensure convergence, such methods, as well as the well-known tâtonnement process

$$p^{k+1} = \pi_{R^n_+}(p^k + \lambda_k q^k), q^k \in E(p^k), \lambda_k > 0. \tag{3.24}$$

need additional assumptions on E (or G). Indeed, the mapping E in (3.20), (3.19) is not in general integrable, hence, such methods may fail even if $G = -E$ is not strictly monotone; e.g., see Example 1.2.1. In particular, process (3.24) is convergent if the following assumption hold:

$$\forall p \in R^n_+ \backslash P^*, \forall g \in G(p) : \langle g, p - p^* \rangle > 0, \tag{3.25}$$

where p^* is any equilibrium point; e.g., see [68, Chapter 5]. Assumption (3.25) can be treated as a variant of the revealed preference condition. Let us consider the problem of finding a point $p^* \in R^n_+$ such that

$$\forall p^* \in R^n_+, \forall g \in G(p) : \langle g, p - p^* \rangle \geq 0, \tag{3.26}$$

i.e., the dual problem to (3.23). It is clear that (3.26) is essentially weaker than (3.25), however, all the CR methods described in Chapters 1 and 2 are convergent if there is a solution to problem (3.26). On the other hand, these methods are much more simpler than the simplicial type methods. Therefore, we intend to consider possible applications of CR methods to several economic equilibrium models. These models have been well documented in literature. In fact, if (3.26) (or (3.25)) holds, it is suitable to compute economic equilibrium in these models with the help of CR methods.

3.2.2 Economic Equilibrium Models with Monotone Mappings

We first consider the case where the budgets of comsumers are fixed. As to producers, we suppose that the supply function $S_j(p)$ for the jth producer (sector) is defined by

$$S_j(p) = \text{Argmax}\{\langle p, y^j \rangle \mid y^j \in Y^j\}, \tag{3.27}$$

where $Y^j \subseteq R^n$ is the production set. In other words, the supply of the jth producer maximizes its profit subject to technology constraints. The mapping S_j in (3.27) is in general multivalued, however, it is easy to see that S_j is monotone, if it has nonempty values. Indeed, take any p', p'' and $s' \in S_j(p')$, $s'' \in S_j(p'')$. Then, by definition,

$$\langle p', s' - s'' \rangle \geq 0$$

and

$$\langle p'', s'' - s' \rangle \geq 0.$$

Adding these inequalities gives

$$\langle p' - p'', s' - s'' \rangle \geq 0$$

i.e., S_j is monotone. Next, since S_j is monotone for each j, so is S in (3.19). Thus, we have proved the following.

Proposition 3.2.1. *If the mapping S, defined by (3.19) and (3.27), has nonempty values, then it is monotone.*

Note that it is usually assumed that Y_j is convex and closed. Hence, in order to guarantee for S_j to have nonempty values it suffices to suppose the boundedness of Y_j.

We now consider monotonicity properties of demand. Suppose that, for each ith consumer, its wealth w_i is fixed. Then, the demand function $D_i(p)$ is defined by

$$D_i(p) = \mathrm{Argmax}\{\varphi_i(x, p) \mid \langle p, x \rangle \leq w_i, \ x^i \in X_i\}, \tag{3.28}$$

where $X_i \subseteq R^n$ is the consumption set, φ_i is the utility function.

Proposition 3.2.2. *(e.g., see [177, Lemma 3.2.1]) Suppose that:*
(a) X_i is a convex cone;
(b) $\varphi_i(x, p) \equiv \psi_i(x)$ is positively homogenous with degree $\alpha_i > 0$;
(c) there is $x \in X_i$ such that $f_i(x) > 0$;
(d) $w_i > 0$.
Then, it holds that

$$D_i(p) = \mathrm{Argmax}\left\{\frac{w_i}{\alpha_i} \ln \psi_i(x) - \langle p, x \rangle \mid x \in X_i\right\}. \tag{3.29}$$

So, problems in (3.28) and (3.29) turned out to be equivalent. Again, this property implies the monotonicity of $-D_i$. In fact, take any p', p'' and $d' \in D_i(p'), d'' \in D_i(p'')$. Then, by definition,

$$\frac{w_i}{\alpha_i} \ln \psi_i(d') - \langle p', d' \rangle \geq \frac{w_i}{\alpha_i} \ln \psi_i(d'') - \langle p', d'' \rangle$$

and

$$\frac{w_i}{\alpha_i} \ln \psi_i(d'') - \langle p'', d'' \rangle \geq \frac{w_i}{\alpha_i} \ln \psi_i(d') - \langle p'', d' \rangle.$$

Adding these inequalities yields

$$\langle p'' - p', d' - d'' \rangle \geq 0,$$

i.e., the mapping $-D_i$ is monotone. Since for each i, $-D_i$ is monotone, so is $-D$ and we obtain the following.

Proposition 3.2.3. *Suppose that the assumptions of Proposition 3.2.2 hold for each $i \in M_c$ and that the mapping D, defined by (3.19) and (3.28), have nonempty values. Then, $-D$ is monotone.*

Thus, if all the assumptions of Propositions 3.2.1 and 3.2.3 hold, then the mapping G, defined by (3.19), (3.20), and (3.22), is monotone. Therefore, to find equilibrium points we can apply the CR methods from Chapter 2 or the iterative methods from Section 3.1.

There exist other models of demand functions which enjoy monotonicity properties. In particular, in [177, pp.70–72], similar monotonicity properties were established for either concave or separable homogenous cost functions ψ_i, however, the demand function $D(p)$ then need not be integrable. Let us now consider the case where

$$D_i(p) = \text{Argmax}\{\langle c^i, x \rangle - \langle p, x \rangle \mid \langle p, x \rangle \leq w_i, x \in R^n_+\}, \tag{3.30}$$

i.e., the utility functions include the cost of commodities and $X_i = R^n_+$.

Proposition 3.2.4. *[157, Theorem 1.2] If D_i is defined by (3.30), then $-D_i$ is monotone on $\text{int} R^n_+$.*

Note that D_i in (3.30) is in general multivalued as well. Combining Propositions 3.2.1 and 3.2.4, we obtain the following assertion.

Proposition 3.2.5. *Suppose that, for each $i \in M_c$, the mapping D_i is defined by (3.30) and, for each $j \in M_s$, the mapping S_j, defined by (3.27), has nonempty values on R^n_+. Then, the mapping G defined by (3.19), (3.20) and (3.22), is monotone on $\text{int} R^n_+$.*

So, under the assumptions of Proposition 3.2.5, one can find a solution to problem (3.21) (or (3.23)) with the help of the CR methods from Chapter 2 or the iterative methods from Section 3.1.

3.2.3 The Linear Exchange Model

In this model, all the agents are consumers, i.e., $M_s = \emptyset$, the endowments of commodities of the ith consumer being determined as a vector $s^i \in R^n$. It follows that

$$S(p) \equiv S = \left\{ \sum_{i \in M} s^i \right\}.$$

On the other hand, the demand function $D_i(p)$ is defined by

$$D_i(p) = \text{Argmax}\{\langle c^i, x^i \rangle \mid \langle p, x^i \rangle \leq \langle p, s^i \rangle, \ x^i \in R^n_+\}, \qquad (3.31)$$

where $c^i \in R^n_+$. According to (3.19), (3.20), set

$$D(p) = \sum_{i \in M} D_i(p)$$

and

$$E(p) = D(p) - S. \qquad (3.32)$$

Let us consider properties of the excess demand function $E(p)$ in (3.32). It is clear that D is in general multivalued and also positively homogenous of degree 0, i.e.,

$$D(\alpha p) = D(p) \quad \forall \alpha > 0.$$

It follows that the mapping G defined by (3.22) and (3.32) is not monotone on R^n_+; e.g., see [177]. Moreover, even if some vectors s^i in (3.31) have zero components, then condition (3.26), as a rule, is satisfied; see [181]. Therefore, we can find a solution of this model with the help of Method 2.2 or 2.3.

To illustrate this approach, we give results of numerical experiments with the linear exchange models. First we apply Method 2.2 to solve test problems and set

$$h(u) = \sum_{j=1}^{n} \max\{-u_j, 0\}.$$

The sequences $\{\gamma_j\}$, $\{\varepsilon_l\}$ and $\{\eta_l\}$ used in Method 2.2 were given by

$$\gamma_j = \gamma = 1.5, \varepsilon_{l+1} = \nu\varepsilon_l, \eta_{l+1} = \nu\eta_l;$$

where $\nu = 0.5$. The parameter θ is chosen to be 0.5. We used the following accuracy measure

$$\text{crit} = \min\{\delta_1, \delta_2\},$$

where δ_1 is chosen to be η_l and δ_2 is chosen to be $\|g^k\|$.

We implemented Method 2.2 in PASCAL, with double precision arithmetic. Each test problem is determined by two matrices C and S with the columns c^i and s^i, respectively. For each problem we report: the value of crit, the number of inner steps (st), the number of iterations (it), and the current iteration point.

Example 3.2.1. Set $m = 4$, $n = 2$,

$$C = \begin{pmatrix} 1 & 1 & 2 & 2 \\ 1 & 2 & 1 & 5 \end{pmatrix}, S = \begin{pmatrix} 5 & 1 & 2 & 2 \\ 2 & 1 & 1 & 4 \end{pmatrix}.$$

The solution set of this problem is the following:

$$P^* = \{p \in R^2 \mid p = (\alpha, \alpha)^T, \quad \alpha > 0\}.$$

We used the starting point $p^0 = (1,5)^T$ and the following initial values: $\varepsilon_0 = 2$, $\eta_0 = 5$.

Example 3.2.2. Set $m = 7$, $n = 4$,

$$C = \begin{pmatrix} 1 & 2 & 3 & 2 & 2 & 3 & 1 \\ 1 & 1 & 2 & 2 & 3 & 4 & 3 \\ 2 & 1 & 1 & 1 & 1 & 3 & 4 \\ 3 & 1 & 1 & 5 & 1 & 1 & 3 \end{pmatrix}, S = \begin{pmatrix} 5 & 3 & 1 & 2 & 2 & 1 & 2 \\ 2 & 1 & 1 & 2 & 1 & 3 & 4 \\ 1 & 1 & 4 & 1 & 2 & 1 & 3 \\ 2 & 2 & 3 & 3 & 2 & 2 & 4 \end{pmatrix}.$$

The solution set of this problem is the following:

$$P^* = \{p \in R^4 \mid p = (\alpha, \alpha, \alpha, \alpha)^T, \quad \alpha > 0\}.$$

We used the starting point $p^0 = (1, 2, 3, 4)^T$ and the following initial values: $\varepsilon_0 = 15$, $\eta_0 = 25$.

Example 3.2.3. Set $m = 5$, $n = 2$,

$$C = \begin{pmatrix} 5 & 2 & 3 & 2 & 2 \\ 2 & 1 & 2 & 2 & 1 \end{pmatrix}, S = \begin{pmatrix} 3 & 3 & 1 & 1 & 4 \\ 2 & 2 & 3 & 1 & 2 \end{pmatrix}.$$

The solution set of this problem is the following:

$$P^* = \{p \in R^2 \mid p = (2\alpha, \alpha)^T, \quad \alpha > 0\}.$$

Note that the mapping D is set-valued on P^*. We used the starting point $p^0 = (1, 5)^T$ and the following initial values: $\alpha_0 = 15$, $\eta_0 = 25$.

The results of numerical experiments for Examples 3.2.1 – 3.2.3 are shown in Table 3.1.

Table 3.1.

Example 3.2.1	crit	2.5	0.0006
	st	3	14
	it	1	10
	point	–	$(3.535, 3.536)^T$
Example 3.2.2	crit	3.12	0.00076
	st	21	57
	it	1	9
	point	–	$(2.656, 2.656, 2.656, 2.656)^T$
Example 3.2.3	crit	19.14	0.1
	st	10	20
	it	1	4
	point	–	$(4.15, 2.04)^T$

For the following test problem, we tested Method 2.2 with various values of the dimensionality.

Example 3.2.4. Set $m = n$,

$$c^i_j = \begin{cases} 1 & \text{if } i = j \text{ or } j = i+1 \le n, \\ 0 & \text{otherwise;} \end{cases}$$

and

$$s^i_j = \begin{cases} \tau_i & \text{if } i \ne j, \\ 0 & \text{if } i = j; \end{cases}$$

where

$$\tau_i = \bar{\tau}_i / \sum_{j=1}^{n} \bar{\tau}_j, \quad i = 1, \ldots, n;$$

$$\bar{\tau}_i = 1.5 - \sin i, \quad i = 1, \ldots, n.$$

Thus, the ith consumer restricts his demands with two commodities (one if $i = n$), including the absent commodity. The solution set of this problem is the following:

$$P^* = \{p \in R^n \mid p = \alpha(\tau_1, \ldots, \tau_n)^T, \quad \alpha > 0\}.$$

The values of ε_0 and η_0 were chosen to be 2 and 5, respectively. The results for two starting points are given in Table 3.2. According to the results, the

Table 3.2.

		n	3	7	10	20	30
$u^0 = (1, \ldots, 1)^T$	st		17	725	1463	169	13
	it		5	72	107	21	1
	crit		0.096	0.08	0.08	0.1	0.09
$u^0 = (1/n, \ldots, 1)^T$	st		3	49	1387	239	39
	it		1	9	86	26	5
	crit		0.09	0.08	0.08	0.1	0.095

middle-dimensional problems turn out to be more difficult for the method, although the convergence is rather fast in general.

Example 3.2.5. Set $m = n$,

$$c^i_j = \begin{cases} 1 & \text{if } i = j, \\ 0 & \text{otherwise;} \end{cases}$$

the values of s^i_j and τ_i are the same as those in Example 3.2.4.

Thus, the ith consumer now intends to purchase only the absent commodity. It follows that the problem has the same solution set

$$P^* = \{p \in R^n \mid p = \alpha(\tau_1, \ldots, \tau_n)^T, \quad \alpha > 0\},$$

however, the mapping E is now single-valued.

Hence, we can apply Methods 1.1 – 1.4 to solve this problem as well. We also implemented these methods in PASCAL, with double precision arithmetic and performed two series of experiments. In both cases, the accuracy measure (crit) was chosen to be the distance to a solution, and the starting point was chosen to be $(1/n, 2/n, \ldots, 1)^T$. In the first series, we performed Method 2.2 with the same values of parameters as those in the previous example. We performed the simplest variants of Methods 1.1 – 1.4, i.e., we used the rule (1.27) with $A_k \equiv I$. We used the following values of parameters:

$$\gamma_k = \gamma = 1.8, \alpha = 0.3, \beta = 0.5, \tilde{\theta} = 1.$$

The results are given in Table 3.3. Thus, the convergence of all the methods

Table 3.3.

Method	n=10		n=20		n=30	
	it	crit	it	crit	it	crit
1.1	10	0.0006	38	0.001	83	0.0005
1.2	11	0.0008	18	0.0006	24	0.0008
1.3	17	0.001	23	0.0009	32	0.0008
1.4	11	0.0008	18	0.0006	24	0.0008
2.2	13	0.0009	13	0.0008	12	0.0007

is rather fast for this problem.

In the second series, we investigate the performance of Methods 1.2 – 1.4 for other choices of T_k. Namely, we used rule (1.27) with A_k being the diagonal part of $\nabla G(u^k)$ (Jacobi) and with A_k being the sum of the diagonal and the lower triangular part of $\nabla G(u^k)$ (overrelaxation). The other parameters of the methods were chosen to be the same as in the first series. The results are given in Table 3.4. According to these results, the convergence of all the

Table 3.4.

Jacobi	n=10		n=20		n=30	
	it	crit	it	crit	it	crit
1.2	8	0.0007	10	0.0005	9	0.001
1.3	12	0.0005	13	0.0009	18	0.0007
1.4	8	0.0007	10	0.0005	9	0.0001
overrelaxation						
1.2	8	0.001	9	0.0008	9	0.0006
1.3	11	0.0009	15	0.0006	16	0.0008
1.4	8	0.001	9	0.0008	9	0.0006

methods is considerably faster than that in the previous series.

3.2.4 The General Equilibrium Model

We now consider an economic equilibrium model which involves production and consumption. Moreover, it employs a linear activity matrix in order to describe the supply function. More precisely, we suppose that $M_s \cap M_c = \emptyset$ and that $M_s = \{j_1, \ldots, j_l\}$, $M_c = \{i_1, \ldots, i_t\}$, $l + t = m$. In accordance with (3.27), the supply function $\tilde{S}_{j_q}$ for the j_qth producer is defined by

$$\tilde{S}_{j_q}(p) = \text{Argmax}\{\langle p, y \rangle \mid y \in Y_{j_q}\}. \tag{3.33}$$

To describe the production sets Y_{j_q} we introduce the $n \times l$ technology matrix A whose columns a^k convert activities of producers into commodities as follows:

$$Y_{j_q} = \{y \in R^n \mid y = z_q a^q, z_q \geq 0\}, \tag{3.34}$$

where z_q is the activity of the j_qth producer. Combining (3.33) and (3.34) gives

$$\tilde{S}_{j_q}(p) = \{y \in R^n \mid y = z_q a^q, z_q \geq 0, z_q (a^q)^T p \geq w(a^q)^T p \quad \forall w \geq 0\}.$$

Replacing this VI with the equivalent CP (see Proposition 1.1.7), we see that

$$\tilde{S}_{j_q}(p) = z_q a^q, \text{ where } z_q \geq 0, (a^q)^T p \leq 0, z^q (a^q)^T p = 0.$$

Thus, the supply function of all the producers is determined as follows

$$\tilde{S}(p) = Az, \text{ where } z \geq 0, -A^T p \geq 0, (-A^T p)^T z = 0. \tag{3.35}$$

Next, the i_qth consumer has endowments of commodities which are described by a vector $b^k \in R^n$ and a demand function $\tilde{D}_{i_q} : R_+^n \to R^n$. For simplicity, set $D_q = \tilde{D}_{i_q}$ for $q = 1, \ldots, t$ and $b = \sum_{q=1}^t b^q$. Then the excess demand function E can be determined as follows:

$$E(p) = -\tilde{S}(p) - b + \sum_{q=1}^t D_q(p).$$

Therefore, the equilibrium prices p^* satisfy the following complementarity conditions:

$$p^* \geq 0, -E(p^*) \geq 0, -E(p^*)^T p = 0. \tag{3.36}$$

Combining (3.35) and (3.36), we now obtain the problem of finding prices $p^* \in R^n$ and activities $z^* \in R^l$ such that

$$p^* \geq 0, z^* \geq 0;$$

$$-A^T p^* \geq 0, Az^* + b - \sum_{q=1}^t D_q(p^*) \geq 0; \tag{3.37}$$

$$(-A^T p^*)^T z^* = 0, (Az^* + b - \sum_{q=1}^t D_q(p^*))^T p^* = 0.$$

It is clear that (3.37) is a complementarity problem. In fact, set $N = n + l$,

$$u = \begin{pmatrix} p \\ z \end{pmatrix} \in R^N$$

and

$$G(u) = \begin{pmatrix} -A^T p \\ Az + b - \sum_{q=1}^t D_q(p). \end{pmatrix}. \tag{3.38}$$

Then (3.37) is equivalent to the problem of finding $u^* \in R^N$ such that

$$u^* \geq 0, G(u^*) \geq 0, G(u^*)^T u^* = 0. \tag{3.39}$$

In turn, due to Proposition 1.1.6 (or 2.1.6), this problem is equivalent to VI (1.1) defined in the space R^N, where $U = R_+^N$. Therefore, CR methods can be in principle applied to find a solution to problem (3.38), (3.39). It should be noted that the demand functions are assumed to obey the Walras law:

$$\langle D_q(p), p \rangle = \langle b^q, p \rangle \quad \text{for all} \quad p \in R_+^n \text{ and } q = 1, \dots, t. \tag{3.40}$$

However, even under condition (3.40), the mapping G in (3.38) is in general non-monotone. Moreover, it is easy to give an example of problem (3.38), (3.39), for which the dual problem (1.2) (or (3.26)) has no solution. Therefore, an iterative sequence being generated by a CR method as well as by any method from Section 3.1 need not converge to a solution of problem (3.38), (3.39). We now consider results of numerical experiments with the well-known Mathiesen example [148] of the general equilibrium model.

Example 3.2.6. Set $n = 3, l = 1, m = 3$,

$$A = (1 - 1 - 1)^T \quad \text{and} \quad b = (0, b_2, b_3)^T \text{ with } b_2 > 0, b_3 > 0.$$

Also, let the demand functions be defined as follows:

$$D_i(p) = h_i(b_2 p_2 + b_3 p_3)/p_i \quad \text{for } i = 1, 2, 3;$$

where $h = (\mu, 1 - \mu, 0)^T$, $0 < \mu < 1$. Therefore, we have $u = (z, p_1, p_2, p_3)^T$,

$$G(u) = \begin{pmatrix} -p_1 + p_2 + p_3 \\ z - \mu(b_2 p_2 + b_3 p_3)/p_1 \\ b_2 - z - (1 - \mu)(b_2 p_2 + b_3 p_3)/p_2 \\ b_3 - z \end{pmatrix}.$$

If we choose the parameters as follows: $b_2 = 5, b_3 = 3, \mu = 0.9$, then problem (3.39) has the solution set

$$U^* = \{u^* \in R^4 \mid u^* = (3, 6\alpha, \alpha, 5\alpha)^T, \alpha > 0\}.$$

If we now take the feasible point $u = (3, 6\alpha, \alpha + 1, 5\alpha + 6)^T$, then

$$\langle G(u), u - u^* \rangle = 1.5 - 0.3\frac{5(\alpha + 1) + 1}{\alpha + 1} = -0.3/(\alpha + 1) < 0,$$

i.e., DVI (1.2) has no solution.

For the above test problem, we tested Methods 1.1 and 1.4 and used rule (1.27) with $A_k \equiv I$. We implemented both the methods in PASCAL, with double precision arithmetic. The parameters of the methods were chosen as follows:

$$\gamma_j = \gamma = 1.8, \alpha = 0.3, \beta = 0.5, \tilde{\theta} = 1.$$

The accuracy measure (crit) was chosen to be the distance to a solution. For each problem, we also report the number of iterations (it). The results for two sets of the parameters b and μ are given in Table 3.5. According to these results, both the methods fail in the case where $\mu = 0.9, b = (0, 5, 3)^T$ and $u^0 = (1, 1, 1, 1)^T$. This fact is not surprising due to the above considerations. Moreover, it was noticed in [148, p. 17] that even fixed-point type algorithms may fail for this problem. In other cases, the rates of convergence for both the

Table 3.5.

Parameters	Starting point	Method 1.1		Method 1.4	
		it	crit	it	crit
$\mu = 0.9$ $b = (0, 5, 3)^T$	$(1, 2, 3, 4)^T$	27	0.0004	24	0.0004
	$(4, 3, 2, 1)^T$	23	0.0009	21	0.0003
	$(1, 1, 1, 1)^T$	101	1.609	101	1.767
$\mu = 0.9$ $b = (0, 1, 0.5)^T$	$(1, 2, 3, 4)^T$	35	0.0004	36	0.0009
	$(4, 3, 2, 1)^T$	43	0.001	51	0.0005
	$(1, 1, 1, 1)^T$	32	0.0008	25	0.0008

methods are close to each other. Therefore, it seems reasonable to consider additional conditions under which the CR methods and methods from Section 3.1 will generate sequences converging to a solution of the general equilibrium model.

3.2.5 The Oligopolistic Equilibrium Model

We now consider the problem of finding equilibria for the case of a few economic agents (producers). More precisely, it is assumed that there are n firms supplying a homogeneous product and that the price p depends on its quantity σ, i.e. $p = p(\sigma)$ is the inverse demand function. Next, the value $h_i(x_i)$ represents the ith firm total cost of supplying x_i units of the product. Naturally, each firm seeks to maximize its own profit by choosing the corresponding production level. Thus, the oligopolistic market equilibrium problem can be formulated as a Nash equilibrium problem in the n-person noncooperative game, where the ith player has the strategy set R_+ and the utility function

$$f_i(x_1, \ldots, x_n) = x_i p \left(\sum_{j=1}^n x_j \right) - h_i(x_i). \tag{3.41}$$

Therefore, a point $x^* = (x_1^*, \ldots, x_n^*)^T \in R_+^n$ is a solution for the oligopolistic market equilibrium problem, if

$$f_i(x_1^*, \ldots, x_{i-1}^*, y_i, x_{i+1}^*, \ldots, x_n^*)$$
$$\leq f_i(x_1^*, \ldots, x_n^*) \quad \forall y_i \in R_+, \ i = 1, \ldots, n. \tag{3.42}$$

Following (2.16) and (2.17), we set

$$\Psi(x, y) = -\sum_{i=1}^{n} f_i(x_1, \ldots, x_{i-1}, y_i, x_{i+1}, \ldots, x_n) \tag{3.43}$$

and

$$\Phi(x, y) = \Psi(x, y) - \Psi(x, x), \tag{3.44}$$

then the Nash equilibrium problem (3.41), (3.42) becomes equivalent to EP (2.12) with $U = R_+^n$. We shall consider this problem under the following assumptions.

(H3.2) *Let $p : R_+ \to R_+$ be twice continuously differentiable and nonincreasing and let the function $\mu_\tau : R_+ \to R_+$, defined by $\mu_\tau(\sigma) = \sigma p(\sigma + \tau)$, be concave for every $\tau \geq 0$. Also, let the functions $h_i : R_+ \to R, i = 1, \ldots, n$ be convex and twice continuously differentiable.*

We use the differentiability assumptions only for simplicity, the rest of the assumptions seems rather reasonable and general. Under these assumptions, by (3.43) and (3.44), the function $\Phi(x, \cdot)$ is convex and twice continuously differentiable. For brevity, set

$$H(x) = (h_1'(x_1), \ldots, h_n'(x_n))^T, e = (1, \ldots, 1)^T \in R^n, \sigma_x = \sum_{i=1}^{n} x_i = \langle x, e \rangle.$$

Applying Theorem 2.1.2 and Proposition 2.1.6, we now obtain the following equivalence result.

Proposition 3.2.6. *The Nash equilibrium problem (3.41), (3.42) is equivalent to VI (1.1) (or CP (1.4)), where $U = R_+^n$ and*

$$G(x) = H(x) - p(\sigma_x)e - p'(\sigma_x)x. \tag{3.45}$$

In order to apply iterative solution methods to problem (3.41), (3.42), we thus need to investigate monotonicity properties of the mapping G in (3.45). First we note that H is clearly monotone and that

$$\nabla G(x) = \nabla H(x) - C(x),$$

where $C(x)$ is defined as follows:

$$\begin{pmatrix} 2p'(\sigma_x) + x_1 p''(\sigma_x) & p'(\sigma_x) + x_1 p''(\sigma_x) & \cdots & p'(\sigma_x) + x_1 p''(\sigma_x) \\ p'(\sigma_x) + x_2 p''(\sigma_x) & 2p'(\sigma_x) + x_2 p''(\sigma_x) & \cdots & p'(\sigma_x) + x_2 p''(\sigma_x) \\ \cdots & \cdots & \cdots & \cdots \\ p'(\sigma_x) + x_n p''(\sigma_x) & p'(\sigma_x) + x_n p''(\sigma_x) & \cdots & 2p'(\sigma_x) + x^n p''(\sigma_x) \end{pmatrix}.$$

Take any $y \in R^n$. Then, taking into account Proposition 1.1.5, we have

$$
\begin{aligned}
\langle \nabla G(x)y, y \rangle \;&=\; \langle \nabla H(x)y, y \rangle - p'(\sigma_x) \sum_{i=1}^{n} y_i^2 \\
&\quad - \sigma_y \sum_{i=1}^{n} (p'(\sigma_x) + x_i p''(\sigma_x)) y_i \\
&\geq\; -p'(\sigma_x) \sum_{i=1}^{n} y_i^2 - p'(\sigma_x)\sigma_y^2 - \sigma_y p''(\sigma_x) \sum_{i=1}^{n} x_i y_i \\
&\geq\; -p'(\sigma_x) \sum_{i=1}^{n} y_i^2 - \sigma_y p''(\sigma_x) \sum_{i=1}^{n} x_i y_i.
\end{aligned}
$$

Hence, if p is affine, then $p''(\sigma_x) = 0$ and we obtain

$$
\langle \nabla G(x)y, y \rangle \geq -p'(\sigma_x)\|y\|^2 \quad \forall y \in R^n.
$$

So, we have proved the following assertion.

Proposition 3.2.7. *Let Hypothesis (H3.2) hold and let p be affine with $p'(\sigma) \neq 0$. Then the mapping G in (3.45) is strongly monotone on R_+^n.*

The general convex case was investigated in [212]. In fact, if the function p is convex, then, by (3.41), (3.43) and (3.44), we see that the function $\Psi(\cdot, \cdot)$ is convex and that the function $\Psi(\cdot, y)$ is concave. Applying now Corollary 2 in [212, p.108], we obtain the following assertion.

Proposition 3.2.8. *Let assumption (H3.2) hold and let p be convex. Then the mapping G in (3.45) is monotone.*

The assumptions of Propositions 3.2.7 and 3.2.8 are satisfied in the linear case originated by A. Cournot [35].

Example 3.2.7. Let the functions p and h_i be affine, i.e.,

$$
\begin{aligned}
p(\sigma) \;&=\; \alpha - \beta\sigma, \alpha \geq 0, \beta > 0; \\
h_i(x_i) \;&=\; \gamma_i x_i + \delta_i, \gamma_i \geq 0, \delta_i \geq 0 \quad \text{for} \quad i = 1,\dots,n.
\end{aligned}
$$

Then,

$$
f_i(x) = x_i(\alpha - \beta\sigma_x) - \gamma_i x_i - \delta_i
$$

and

$$
G(x) = (\gamma_1, \dots, \gamma_n)^T - (\alpha - \beta\sigma_x)e + \beta x. \tag{3.46}
$$

It follows that the Jacobian

$$
\nabla G(x) = \begin{pmatrix}
2\beta & \beta & \dots & \beta \\
\beta & 2\beta & \dots & \beta \\
\dots & \dots & \dots & \dots \\
\beta & \beta & \dots & 2\beta
\end{pmatrix}
$$

is a symmetric and positive definite matrix.

Therefore, the equilibrium problem (3.41), (3.42) is now equivalent to the problem of minimizing the strongly convex quadratic function

$$F(x) = 0.5\beta\sigma_x^2 + 0.5\beta \sum_{i=1}^{n} x_i^2 + \sum_{i=1}^{n} (\gamma_i - \alpha)x_i$$

over the set R_+^n. This problem clearly has a unique solution. Letting $G(x)$ in (3.46) to be equal to zero, we obtain

$$x_i^* = [\alpha - (n+1)\gamma_i + \sigma_\gamma] / [\beta(n+1)] \quad \text{for} \quad i = 1,\ldots,n.$$

If $\alpha \geq (n+1)\gamma_i + \sigma_\gamma$, then x_i^* gives the optimal production level for the ith firm, $i = 1,\ldots,n$. On the other hand, by (3.43) and (3.44), we have

$$
\begin{aligned}
\Phi(x,y) + \Phi(y,x) &= \beta \sum_{i=1}^{n} x_i(\sigma_y - \sigma_x + x_i - y_i) \\
&\quad + \beta \sum_{i=1}^{n} y_i(\sigma_x - \sigma_y + y_i - x_i) \\
&= \beta[\sigma_x(\sigma_y - \sigma_x) + \sigma_y(\sigma_x - \sigma_y) \\
&\quad + \sum_{i=1}^{n} x_i(x_i - y_i) + \sum_{i=1}^{n} y_i(y_i - x_i)] \\
&= \beta\left[\sum_{i=1}^{n}(x_i - y_i)^2 - (\sigma_x - \sigma_y)^2\right] \\
&= -\beta \sum_{1 \leq i \neq j \leq n} (x_i - y_i)(x_j - y_j)
\end{aligned}
$$

for all $x, y \in R_+^n$. If follows that Φ is in general non-monotone, even in case $n = 2$. If suffices to choose $x_1 > y_1$ and $x_2 < y_2$. Therefore, the reverse assertion to Proposition 2.1.17 is not true.

Due to the above results, we can apply CR methods to find oligopolistic market equilibria under rather mild assumptions. We illustrate this assertion on a small numerical example.

Example 3.2.8. [153] Set $m = 5$,

$$h_i(x_i) = \alpha_i x_i + \frac{\beta_i}{\beta_i + 1}\mu_i^{\frac{-1}{\beta_i}} x_i^{\frac{\beta_i + 1}{\beta_i}}$$

for $i = 1,\ldots,5$; $p(\sigma) = (5000/\sigma)^{1/1.1}$. The values of the parameters α_i, β_i and μ_i are given in Table 3.6.

It is easy to see that all the assumptions of Proposition 3.2.8 are satisfied, so that the mapping G is monotone. This example was solved by the method

Table 3.6.

Firm	α_i	β_i	μ_i
1	10	1.2	5
2	8	1.1	5
3	6	1.0	5
4	4	0.9	5
5	2	0.8	5

of [153] with the accuracy 0.001 in six iterations, each iteration involving a solution to an auxiliary nonlinear optimization problem.

We implemented Methods 1.1 and 1.4 in PASCAL, with double precise arithmetic, to solve the same test example. We performed the simplest variants of both the methods, i.e., we used the rule (1.27) with $A_k \equiv I$. The parameters of the methods were chosen as follows:

$$\gamma_k = \gamma = 1.8, \alpha = 0.3, \beta = 0.5, \tilde{\theta} = 1.$$

The accuracy measure (crit) was chosen to be the value of $\|G(x)\|$ and the starting point was chosen to be $(1, 1, 1, 1, 1)^T$. The results are given in Table 3.7, where (it) denotes the number of iterations. Thus, the results demon-

Table 3.7.

Method	it	crit
1.1	22	0.00001
1.4	20	0.000008

strate rather fast convergence of all the CR methods.

3.3 Numerical Experiments with Test Problems

In this section, we give additional results of numerical experiments for some CR methods from Chapters 1 and 2 and compare them with the other methods described in Section 3.1. All the methods were implemented in PASCAL, with double precision arithmetic. The test problems of this section model various difficulties arising in solving variational inequality problems.

3.3.1 Systems of Linear Equations

As indicated in Section 1.2, many projection-type algorithms fail to provide convergence to a solution of VI (1.1) in the case where the main mapping G is not integrable and/or strictly monotone. Motivated by Example 1.2.1, we considered convergence of several iterative methods in the simplest case where $U = R^n$ and $G(u) = Au - b$, A being an $n \times n$ skew-symmetric matrix. It is clear that VI (1.1) then reduces to the system of linear equations

$$G(u) = Au - b \qquad (3.47)$$

(see also (1.3)). More precisely, the test problem can be described as follows.

Example 3.3.1. Let us consider problem (3.47) with $b = 0$ and with the elements of A being defined as follows

$$a_{ij} = \begin{cases} -1 & \text{if } j = n+1-i < i, \\ 1 & \text{if } j = n+1-i > i, \\ 0 & \text{otherwise .} \end{cases}$$

It is clear that problem (3.47) then has the unique solution $u^* = (0, \ldots, 0)^T$ and that A is skew-symmetric. As indicated in Example 1.2.1, the usual method

$$u^{k+1} = u^k - \lambda_k G(u^k)$$

fails to provide convergence to a solution regardless the stepsize rule. Moreover, the descent method from [145] for monotone VI's on compact sets, also falls into the above recurrence since $U = R^n$. Hence, under these assumptions, it is not convergent either. Therefore, we investigate convergence of the following methods: IRM (Method 3.2), AM (Method 3.3), EM (Method 3.4), EGM (Method 3.5), and Method 1.1 (CRM for short).

Here and below δ denotes the prescribed accuracy, $crit_k$ denotes the value of an accuracy measure at the kth iteration.

For all the methods, the starting point u^0 was chosen to be $(1, \ldots, 1)^T$. IRM was implemented in accordance with (3.7), where

$$\alpha_k = \tilde{\alpha}/(k+1)^{1/2}, \ \theta_k = \tilde{\theta}/(k+1)^{1/4}, \ \tilde{\alpha} = 2, \ \tilde{\theta} = 1.$$

In AM, we used the following rule:

$$\alpha_k = \tilde{\alpha}/(k+1), \ \tilde{\alpha} = 1.$$

Other values of the parameters $\tilde{\alpha}$ and $\tilde{\theta}$ in IRM and AM led to slower convergence in our experiments. Next, the parameter θ in EM was chosen to be $\sqrt{n}$, i.e., the condition $u^* \in B(u^0, \theta)$ is then satisfied. EGM was defined by (3.17), (3.18) with

$$\alpha = 0.85, \beta = 0.5, \tilde{\theta} = 1. \tag{3.48}$$

Note that 0.85 is the best value for the parameter α, according to [88]. We implemented the simplest variant of Method 1.1 which uses rule (1.27) with $A_k \equiv I$. We used the following values of its parameters:

$$\gamma_k = \gamma = 1, \alpha = 0.3, \beta = 0.5, \tilde{\theta} = 1.$$

The accuracy measure (crit) for all the methods was chosen to be the distance to a solution. The results are given in Table 3.8, where we report the number of iterations which was performed by a method to obtain the solution u^* with the accuracy $\delta = 0.001$. In accordance with the results, EGM and CRM

Table 3.8.

n	EM	AM	IRM	EGM	CRM
2	54	$\text{crit}_{350} > 0.38$	135	27	21
10	$\text{crit}_{350} > 0.38$	–	159	30	24
50	$\text{crit}_{200} > 0.14$	–	185	32	26
100	–	–	197	34	27

showed essentially faster convergence than EM, AM and IRM, the results of CRM being better than those of EGM in all experiments. It should be also noted that the number of iterations for IRM, EGM and CRM grow rather slow in comparison with the value of n.

3.3.2 Linear Complementarity Problems

Many difficulties arising in solving VI's can be modeled by means of a linear complementarity problem (LCP). Recall that LCP consists in finding an element u^* such that

$$u^* \in R^n_+, \ Mu^* + q \in R^n_+, \ \langle u^*, Mu^* + q \rangle = 0, \tag{3.49}$$

where M is an $n \times n$ matrix, $q \in R^n$. According to Proposition 1.1.6, it is equivalent to VI (1.1) with $U = R^n_+$, $G(u) = Mu + q$. We compared the performance of Method 1.1, Method 1.2 and EGM on two test LCP's.

Example 3.3.2. [203]. Set $M = CC^T$, $q = Mu^* + d$, where

$$u_j^* = \begin{cases} 0 & \text{if } 1 \leq j \leq n/2, \\ 7.5 & \text{if } n/2 < j \leq n; \end{cases} \quad d_j = \begin{cases} 5 & \text{if } 1 \leq j \leq n/4, \\ 0 & \text{if } n/4 < j \leq n; \end{cases}$$

the elements of the matrix C are defined by

$$c_{ij} = 5(i-j)/n.$$

Thus, we have obtained LCP with the ill-conditioned symmetric matrix M, the point u^* being a solution to this problem. To solve this problem, we implemented Methods 1.1 and 1.2 with the modification (1.73), we using the simplest variant (1.27) with $A_k \equiv I$. In all experiments we used the starting point $u^0 = (0,\dots,0)^T$ and the accuracy $\delta = 10^{-5}$. To estimate the accuracy, we chose the value

$$\text{crit} = \|u - \pi_{R_+^n}(u - \tilde{\theta}G(u))\|. \tag{3.50}$$

The parameters of both the methods were defined as follows:

$$\alpha = 0.3, \beta = 0.5, \tilde{\theta} = 1. \tag{3.51}$$

The results for various values of $\gamma_k = \gamma$ with $n = 10$ are given in Table 3.9. Next, we applied EGM defined by (3.17), (3.18) and (3.48) and the same

Table 3.9. $n = 10, \delta = 10^{-5}$

γ	Method 1.1	Method 1.2
1.0	14	45
1.5	71	27
1.8	136	20

variant of Method 1.1 for the cases of $n = 50$ and 100. The results are given in Table 3.10.

Table 3.10.

n	Method 1.1			EGM
	$\gamma = 1.0$	$\gamma = 1.5$	$\gamma = 1.8$	
50	95	56	14	109
100	–	$\text{crit}_{60} < 10^{-3}$	14	121

Example 3.3.3. [71]. We consider problem (3.49), where $b = (-1,\dots,-1)^T$, the elements of M are defined by

$$m_{ij} = \begin{cases} 2 & \text{if } i < j, \\ 1 & \text{if } i = j, \\ 0 & \text{if } i > j. \end{cases}$$

This problem has the unique solution $u^* = (0,\dots,0,1)^T$.

It is well known as a hard problem for several finite methods, e.g. the number of iterations of the Lemke method [134] being applied to this problem grows in an exponential rate with respect to n.

We also implemented Methods 1.1 and 1.2 with the modification (1.73) and used the simplest variant (1.27) with $A_k \equiv I$. The parameters were defined by (3.51). In all experiments, we used the starting point $u^0 = (0, \ldots, 0)^T$ and the accuracy measure (crit) was chosen to be the distance to a solution. We used the accuracy $\delta = 10^{-3}$ for this problem. The results for various values of $\gamma_k = \gamma$ with $n = 10$ are given in Table 3.11.

Table 3.11. $n = 10, \delta = 10^{-3}$

γ	Method 1.1	Method 1.2
1.0	19	19
1.5	12	11
1.8	11	9

Thus, the best value of the parameter γ for both the methods is 1.8. Afterwards we implemented the same variants of Methods 1.1 and 1.2 for the cases $n = 50$ and 100, γ being chosen to be equal to 1.8. Besides, we implemented EGM defined by (3.17), (3.18) and (3.48) for the same problems. The results are given in Table 3.12.

Table 3.12. $\gamma = 1.8, \delta = 10^{-3}$

n	Method 1.1	Method 1.2	EGM
50	20	16	109
100	37	19	121

Thus, the convergence of the CR methods on both the test problems is rather fast. In particular, it turns out to be faster essentially than that of EGM.

3.3.3 Linear Variational Inequalities

We now consider VI (1.1) where the underlying mapping G is affine and the feasible set is defined by affine functions. This problem, for example, arises in matrix game theory. Recall that a point $(x^*, y^*)^T \in S_+^m \times S_+^l$ is a solution of a matrix game, if

$$x^T A y^* \leq (x^*)^T A y^* \leq (x^*)^T A y \quad \forall x \in S_+^m, \forall y \in S_+^l, \tag{3.52}$$

where A is an $m \times l$ matrix, S_+^m is the standard simplex in R^m, i.e.,

$$S_+^m = \{x \in R_+^m \mid \sum_{i=1}^m x_i = 1\},$$

and S_+^l is defined analogously. Thus, we have a particular case of a 2-person non-cooperative game, where

$$f_1(x,y) = f(x,y) = x^T Ay, \, f_2(x,y) = -f(x,y).$$

This problem is clearly equivalent to EP (2.12) where

$$\Phi(u,v) = -(x')^T Ay + x^T Ay, \, u = (x,y), \, v = (x',y'),$$

and $U = S_+^m \times S_+^l$. Since $\Phi(u,\cdot)$ is convex and differentiable, from Theorem 2.1.2 it follows that our problem (3.52) is equivalent to VI (1.1), where

$$G(u) = \nabla\Phi_v(u,v)|_{v=u} = Cu, \qquad (3.53)$$

with

$$C = \begin{pmatrix} 0 & -A \\ A^T & 0 \end{pmatrix}.$$

Due to Proposition 2.1.17, the mapping G in (3.53) must be monotone. However, we see that G is not strictly monotone since it is determined by the skew-symmetric matrix C.

To solve this problem we implemented Methods 1.1 and 1.2 with modification (1.73) and EGM. We used the simplest variant (1.27) with $A_k \equiv I$ of Methods 1.1 and 1.2. The parameters of both methods were defined by (3.51), also we set $\gamma_k = \gamma \in (0,2)$. We used the variant (3.17), (3.18) and (3.48) of EGM.

Example 3.3.4. Set $m = 3, l = 4,$

$$A = \begin{pmatrix} 3 & 6 & 1 & 4 \\ 5 & 2 & 4 & 2 \\ 1 & 4 & 3 & 5 \end{pmatrix}.$$

This problem has the unique solution

$$u^* = (0.125, 0.5, 0.375, 0.0833, 0.4167, 0.5, 0)^T.$$

We used the starting point $u^0 = (1,0,0,1,0,0,0)^T$ and the accuracy measure (crit) was chosen to be the distance to a solution. The results for the accuracy $\delta = 10^{-3}$ are given in Table 3.13.

Example 3.3.5. [46, Example 2.2] Set $m = l,$

$$C(x) = \begin{pmatrix} p & -t_1 & -t_1 & \cdots & -t_1 \\ -t_2 & p & -t_2 & \cdots & -t_2 \\ \cdots & \cdots & \cdots & \cdots & \cdots \\ -t_m & -t_m & -t_m & \cdots & p \end{pmatrix},$$

Table 3.13. $\delta = 10^{-3}$

	Method 1.1 $\gamma = 1.9$	Method 1.2 $\gamma = 1.9$	EGM
it	77	77	93

where $p > 0$, $t_i \geq 0$ for $i = 1, \ldots, m$. This problem has the unique solution $u^* = (x^*, y^*)^T$, where

$$
\begin{aligned}
x_i^* &= (h_i s_h)^{-1}, \quad i = 1, \ldots, m; \\
y_j^* &= (s_h - n + 1)/x_j^*, \quad i = 1, \ldots, m; \\
h_i &= p + t_i, \ s_h = \sum_{i=1}^{m} h_i^{-1}.
\end{aligned}
$$

We used the starting point $u^0 = (x^0, y^0)^T$, where $x^0 = (1, 0, \ldots, 0)^T$ and $y^0 = (1, 0, 0, \ldots, 0)^T$. The accuracy measure (crit) was chosen to be the value given by (3.50). The results for $n = 3$ and the accuracy $\delta = 10^{-3}$ are given in Table 3.14.

Table 3.14. $\delta = 10^{-3}$

	Method 1.1 $\gamma = 1.8$	Method 1.2 $\gamma = 1.8$	EGM
it	42	42	93

Thus, according to Tables 3.13 and 3.14, the convergence rates of all the CR methods are close to each other.

3.3.4 Nonlinear Complementarity Problems

We now present the results of numerical experiments with nonlinear problems. First we consider the nonlinear CP (1.4).

Example 3.3.6. Set $n = 1$, $U = R_+^1$, $G(u) = (u - 1)^2 - 1.01$.

Note that the mapping G is pseudomonotone, but it is not monotone. This problem, which is due to S. Billups (see [19]), seems too simple, nevertheless, it turned out to be a destroying example for most existing descent type solution methods for variational inequality and complementarity problems. In Table 3.15, we give the result of solving this problem by Method 1.1 with the rules (1.27) and $A_k \equiv I$. The method used the starting point $u^0 = 0$ and the parameter $\gamma_k = \gamma = 1.8$, the other parameters were defined by (3.51). The method terminated with the accuracy $|G(u)| < 6 \cdot 10^{-6}$. Method 1.4

Table 3.15.

it	point	it	point
0	0.00000	5	1.95325
1	0.01800	6	1.99884
2	0.10022	7	2.00438
3	0.46092	8	2.00500
4	1.75582		

showed the same results with this problem.

Due to Corollary 1.1.2 and Proposition 1.1.10, the convex optimization problem

$$\min \rightarrow \{f_0(x) \mid f_i(x) \leq 0 \quad i = 1, \ldots, m\}$$

where $f_i : R^l \rightarrow R$, $i = 0, \ldots, m$ are convex differentiable functions, reduces to CP (1.4), where

$$
\begin{aligned}
U &= R^l \times R^m_+, \\
G(u) &= \begin{pmatrix} \nabla f_0(x) + \sum_{i=1}^m y_i \nabla f_i(x) \\ -f(x) \end{pmatrix}, \\
f(x) &= (f_1(x), \ldots, f_m(x))^T, u = (x, y)^T.
\end{aligned}
$$

with G being monotone (see (1.17)). Note that the new problem has only simple constraints. Although G is not in general strictly monotone, we can apply CR methods to solve this problem. Namely, we implemented Method 1.1 with (1.27) and $A_k \equiv I$.

Example 3.3.7. Set $l = m = 1$, $f_0(x) = x^2$, $h_1(x) = x^2 - 4x + 3$. This problem has the unique solution $u^* = (1, 1)^T$.

We used the following values of parameters: $\alpha = 0.5, \beta = 0.5, \tilde{\theta} = 1$ and the starting point $u^0 = (0, 0)^T$. The results for various values of $\gamma_k = \gamma$ are given in Table 3.16.

Table 3.16.

γ	it	point	it	point	it	point
1.0	0	$(0,0)^T$	10	$(0.897, 0.724)^T$	40	$(0.996, 0.990)^T$
1.5	0	$(0,0)^T$	10	$(0.959, 0.888)^T$	25	$(0.996, 0.989)^T$
1.8	0	$(0,0)^T$	10	$(0.939, 1.000)^T$	23	$(1.003, 1.000)^T$

Example 3.3.8. (Rosen-Suzuki). Set $l = 4, m = 3$,

$$
\begin{aligned}
f_0(x) &= x_1^2 + x_2^2 + 2x_3^2 + x_4^2 - 5x_1 - 5x_2 - 21x_3 + 7x_4, \\
f_1(x) &= x_1^2 + x_2^2 + 2x_3^2 + x_4^2 + x_1 - x_2 + x_3 - x_4 - 8, \\
f_2(x) &= x_1^2 + x_2^2 + x_3^2 + 2x_1 - x_2 - x_4 - 5, \\
f_3(x) &= x_1^2 + 2x_2^2 + x_3^2 + 2x_4^2 - x_1 - x_4 - 10.
\end{aligned}
$$

This problem has the unique solution $u^* = (0, 1, 2, -1, 1, 2, 0)^T$.

We used the modification (1.73). The parameters were chosen to be the following: $\gamma_k = \gamma = 1.8$, $\beta = 0.5$, $\tilde{\theta} = 1$. We chose the following accuracy measure

$$
\text{crit} = \min\{\|u^k - \pi_U(u^k - \tilde{\theta}G(u^k))\|, \|g^k\|\},
$$

where g^k is defined by (1.73). The results for various values of α and the starting point $u^0 = (0, \ldots, 0)^T$ are given in Table 3.17.

Table 3.17.

it	$\alpha = 0.1$	$\alpha = 0.3$	$\alpha = 0.5$
1	23.238	23.238	23.238
10	0.187	0.161	0.368
20	0.050	0.078	0.080
30	0.038	0.028	0.037
39	0.009	–	–
41	–	0.007	–
44	–	–	0.009

In accordance with the results, the closer the values of α and γ approximate 0 and 2, respectively, the faster Method 1.1 converges to a solution.

3.3.5 Nonlinear Variational Inequalities

We now consider an example of VI (2.1) in the case where the mapping G is nonlinear and the feasible set U is defined by nonlinear convex constraints.

Example 3.3.9. Set

$$
G(u) = ([1 + |u_n|^3]u_1, [1 + |u_{n-1}|^3]u_2, \ldots, [1 + |u_1|^3]u_n)^T.
$$

Let U be defined by (2.42), where

$$
\begin{aligned}
h(u) &= \max\{h_1(u), h_2(u)\}, \\
h_1(u) &= (u_1 - 3)^2 + \sum_{i=2}^{n} u_i^2 - 9, \\
h_2(u) &= (u_2 + 1)^2 + \sum_{i=2}^{n} u_i^2 - 25.
\end{aligned}
$$

It is easy to see that $U^* = \{(0,\ldots,0)^T\}$ and that $U^d = U^*$. However, G is not quasimonotone on U, as one can see by considering

$$u = (1, 1/\sqrt{n-2}, \ldots, 1/\sqrt{n-2}, 2)^T$$

and

$$v = (2, 1/\sqrt{n-2}, \ldots, 1/\sqrt{n-2}, 1)^T.$$

To solve this problem we implemented Method 2.2 and the exact version of Method 2.3 where $\delta_k = 0$ for all k. In all cases, we used the rule (2.73) where

$$\varepsilon_1 = \sqrt{2}, \eta_1 = 2\sqrt{2}, \gamma_j = \gamma = 1.5, \theta = 0.5, \nu = 0.5.$$

To implement the operator P_k in Method 2.2 we used Procedure 4.1. Also, the accuracy measure (crit) was chosen to be the distance to the solution u^*.

The computational results for the accuracy $\delta = 0.001$ are shown in Table 3.18, where $\tilde{u}^0$ is the starting point, $kd(kn)$ is the number of descent (null) steps, and $id(ip)$ is the number of iterations in the auxiliary procedure $\tilde{D}_k$ (Procedure P_k). Recall that each inner step of the methods includes the computation of an element of G (or ∂h) at the corresponding point. Therefore, the value $id + ip$ gives the total number of such computations.

Table 3.18.

Method 2.2					
$\tilde{u}^0$	n	kd	kn	id	ip
$(1,3)^T$	2	12	12	27	3
$(5,1)^T$		14	11	33	1
$(1,\ldots,1,3)^T$	5	13	11	28	1
$(5,1,\ldots,1)^T$		13	12	28	1
$(1,\ldots,1,3)^T$	10	12	11	26	1
$(5,1,\ldots,1)^T$		14	12	30	1
$(1,\ldots,1,3)^T$	20	12	11	24	1
$(5,1,\ldots,1)^T$		11	11	23	1
Method 2.3					
$\tilde{u}^0$	n	kd	kn	id	
$(1,3)^T$	2	18	12	54	
$(5,1)^T$		19	11	53	
$(1,\ldots,1,3)^T$	5	14	11	34	
$(5,1,\ldots,1)^T$		19	11	49	
$(1,\ldots,1,3)^T$	10	14	12	30	
$(5,1,\ldots,1)^T$		14	11	30	
$(1,\ldots,1,3)^T$	20	15	12	32	
$(5,1,\ldots,1)^T$		15	12	34	

The computational results show that the behavior of both methods is rather stable. In particular, the dimensionality does not seem to seriously affect the performance of the methods.

4. Auxiliary Results

In this chapter, we give some results which either were formulated in the preceding chapters or clarify the origin of some methods which were used as parts of CR methods. We also describe finite algorithms which can be viewed as some realizations of feasible quasi-nonexpansive mappings.

4.1 Feasible Quasi-Nonexpansive Mappings

By construction, the basic schemes for CR methods involve feasible quasi-nonexpansive mappings P_k, see Definition 1.2.2. Given a convex closed set U in R^n, one can use the projection mapping $\pi_U(\cdot)$ as P_k. Indeed, due to Proposition 1.2.1, $\pi_U(\cdot)$ is a feasible quasi-nonexpansive mapping with respect to the set U.

Recall that we denote by $\mathcal{F}(U)$ the class of feasible quasi-nonexpansive mappings with respect to the set U. However, if the feasible set U is defined by

$$U = \{x \in H \mid h_i(x) \leq 0 \quad i = 1, \ldots, m\}, \tag{4.1}$$

where $h_i : R^n \to R$ is a convex, but not necessarily affine function, for each $i = 1, \ldots, m$, then the projection onto U cannot be in general found by a finite algorithm. Therefore, in order to make CR methods implementable in such a case, we have to give finite procedures whose final point p is in U and yields the following nonexpansive property:

$$\|p - u\| \leq \|v - u\| \qquad \forall u \in U, \tag{4.2}$$

where v is a given starting point. Such procedures are described in this section.

4.1.1 Exterior Point Algorithms

We first describe algorithms which generate an infeasible iteration sequence and terminate with a point $p \in U$ satisfying (4.2).

Hypothesis (H4.1)

(a)   *Given a point $v \in R^n$ and the set U defined by (4.1), where $h_i : R^n \to R$ is convex for each $i = 1, \ldots, m$;*

(b)   *the Slater regularity condition is satisfied, i.e., there is a point $\bar{u}$ such that $h_i(\bar{u}) < 0$ for all $i = 1, \ldots, m$.*

The first procedure for finding a point of U can be described as follows. Set

$$I_+(u) = \{ i \mid 1 \leq i \leq m, \quad h_i(u) > 0 \}.$$

Procedure 4.1. *Data:* A point $v \in R^n$.
   Output: A point p.
   Step 0: Set $w^0 := v$, $j := 0$.
   Step 1: If $w^j \in U$, set $p := w^j$ and stop.
   Step 2: Choose $i(j) \in I_+(w^j)$, $q^j \in \partial h_{i(j)}(w^j)$, set

$$w^{j+1} := w^j - 2h_{i(j)}(w^j)q^j / \|q^j\|^2, \qquad (4.3)$$

$j := j + 1$ and go to Step 1.

Procedure 4.1 is a variant of the reflection method [131] generalizing the relaxation method for solving linear inequalities. Indeed, the points w^j and w^{j+1} in (4.3) are symmetric with respect to the hyperplane

$$H_j = \{ u \in R^n \mid \langle q^j, u - w^j \rangle = -h_{i(j)}(w^j) \},$$

which separates w^j and U. This property leads to the following useful assertions.

Lemma 4.1.1. *If Procedure 4.1 terminates, then $p \in U$ and (4.2) holds.*

Proof. First we note that, by construction, $p \in U$. Next, if $v \in U$, then $p = v$ and (4.2) evidently holds. Let $v \notin U$. Fix $u \in U$. Then, using the subgradient property gives

$$-h_{i(j)}(w^j) \geq h_{i(j)}(u) - h_{i(j)}(w^j) \geq \langle q^j, u - w^j \rangle$$

for each $j = 0, 1, \ldots$ Hence

$$
\begin{aligned}
\|w^{j+1} - u\|^2 &= \|w^j - u\|^2 - 4h_{i(j)}(w^j)\langle q^j, w^j - u \rangle / \|q^j\|^2 \\
&\quad + (2|h_{i(j)}(w^j)|/\|q^j\|)^2 \\
&\leq \|w^j - u\|^2
\end{aligned}
$$

for $j = 0, 1, \ldots$ These inequalities give (4.2).   □

Proposition 4.1.1. *[131, Theorem 2] Procedure 4.1 is finite.*

We now give another finite procedure in the differentiable case, which is based on the linearization method [183, Chapter 3]. Set

$$h(u) = \max_{i=1,\ldots,m} h_i(u),$$
$$\tilde{I}_\delta(u) = \{i \mid 1 \le i \le m, \quad h_i(u) \ge h(u) - \delta\}.$$

Procedure 4.2. *Data:* A point $v \in R^n$.
Output: A point p.
Parameter: A number $\varepsilon \in (0, 1)$.
Step 0: Set $w^0 := v$, $j := 0$.
Step 1: If $w^j \in U$, set $p := w^j$ and stop.
Step 2: Find p^j as the solution of the problem:

$$\min \to \|p\|^2$$

subject to

$$h_i(w^j) + \langle \nabla h_i(w^j), p \rangle \le 0 \qquad i \in \tilde{I}_\delta(w^j).$$

Step 3: Choose $\gamma_j \in [1 + \varepsilon, 2]$, set $w^{j+1} := w^j + \gamma_j p^j$, $j := j + 1$ and go to Step 1.

If all the functions h_i, $i = 1, \ldots, m$ are differentiable, then it was shown in [132] that the assertions of Lemma 4.1.1 and Proposition 4.1.1 are true for Procedure 4.2 as well.

4.1.2 Feasible Point Algorithms

Another way to obtain a point of U satisfying (4.2) consists in generating a sequence of feasible points. Fix a point $v \in R^n$. First we give an equivalent form of condition (4.2), which is more suitable for verification. For each pair of points $u, p \in R^n$, we set

$$\theta(p, u) = \langle p - v, u - 0.5(p + v) \rangle.$$

Lemma 4.1.2. *Let Q be a nonempty, closed and convex set in R^n. Then:*
(i) the inequality

$$\theta(p, u) \ge 0 \quad \forall u \in Q \tag{4.4}$$

is equivalent to

$$\|p - u\| \le \|v - u\| \quad \forall u \in Q;$$

(ii) it holds that

$$\theta(\pi_Q(v), u) \ge 0.5\|\pi_Q(v) - v\|^2 \quad \forall u \in Q. \tag{4.5}$$

Proof. Set $z = 0.5(p + v)$ and take an arbitrary point $u \in Q$. Then we have

$$
\begin{aligned}
\|p - u\|^2 - \|v - u\|^2 &= \|p - v\|^2 + 2\langle p - v, v - u \rangle \\
&= \|p - v\|^2 + 2(\langle p - v, v - z \rangle + \langle p - v, z - u \rangle) \\
&= 2\langle p - v, z - u \rangle = 2\theta(p, u),
\end{aligned}
$$

and the result of (i) follows. In case (ii), from Proposition 1.2.1 (ii) it follows that

$$
\langle \pi_Q(v) - v, u - \pi_Q(v) \rangle \geq 0 \quad \forall u \in Q.
$$

Using this inequality gives

$$
\begin{aligned}
\theta(\pi_Q(v), u) &= \langle \pi_Q(v) - v, u - 0.5(\pi_Q(v) + v) \rangle \\
&= \langle \pi_Q(v) - v, u - \pi_Q(v) \rangle \\
&\quad + \langle \pi_Q(v) - v, 0.5(\pi_Q(v) - v) \rangle \\
&\geq 0.5\|\pi_Q(v) - v\|^2,
\end{aligned}
$$

i.e., (4.5) holds. $\qquad\square$

Condition (4.4) with $Q = U$ takes the advantage over (4.2) since $\theta(p, u)$ is linear in u. Moreover, (4.5) shows that (4.2) (or (4.4) with $Q = U$) can be satisfied in some neighborhood of $\pi_U(v)$. However, it is not easy to verify even the condition (4.4) with $Q = U$ since U is defined by nonlinear functions. In order to overcome this drawback we intend to replace U with a "simpler" set, in the sense that the latter can be defined by affine constraints. For the sake of simplicity, we restrict ourself with the smooth case, i.e., throughout this subsection we suppose that Hypothesis (H4.1) holds and that $h_i : R^n \to R$ is continuously differentiable for each $i = 1, \ldots, m$. Also, since the case $v \in U$ is trivial, we suppose that $v \notin U$.

For each point $w \in R^n$ and for each number ε, we set

$$
\begin{aligned}
I_\varepsilon(w) &= \{i \mid 1 \leq i \leq m, h_i(w) \geq -\varepsilon\}, \\
U_\varepsilon(w) &= \{u \in R^n \mid h_i(w) + \langle \nabla h_i(w), u - w \rangle \leq 0 \quad \forall i \in I_\varepsilon(w)\}.
\end{aligned}
$$

Lemma 4.1.3. *Set $p^* = \pi_U(v)$. Then:*
(i) p^* is the projection of v onto $U_0(p^*)$;
(ii) it holds that

$$
\theta(p^*, u) \geq 0.5\|p^* - v\|^2 \quad \forall u \in U_0(p^*). \tag{4.6}
$$

Proof. By definition, p^* is a unique solution to the optimization problem:

$$
\min \to \{f(u) \mid u \in U\}, \quad \text{where} \quad f(u) = 0.5\|u - v\|^2. \tag{4.7}
$$

Using the Karush-Kuhn-Tucker theorem (e.g. see [225, Section 2.6, Theorem 6]), we see that there exist nonnegative numbers $y_i, i \in I_0(p^*)$ such that

$$\nabla f(p^*) + \sum_{i \in I_0(p^*)} y_i \nabla h_i(p^*) = 0.$$

In turn, due to the same theorem, this relation implies that p^* is a unique solution to the following optimization problem:

$$\min \rightarrow \{f(u) \mid u \in U_0(p^*)\},$$

i.e., p^* is the projection of v onto $U_0(p^*)$. Hence, part (i) holds. Part (ii) follows immediately from (i) and Lemma 4.1.2 (ii). $\square$

Let D be a subset of R^n. Then, for each $w \in R^n$ and for each ε, we set

$$D_\varepsilon(w) = D \bigcap U_\varepsilon(w).$$

We will say that a sequence $\{w^j\}$ is feasible, if $w^j \in U$ for all j. The following assertion gives a basis for constructing various finite feasible point algorithms.

Proposition 4.1.2. *Suppose that there exists a convex polyhedron D such that $D \supseteq U$. Let $\{w^j\}$ be a feasible sequence converging to $p^* = \pi_U(v)$. Then, for each $\varepsilon > 0$, there exists a number t such that*

$$\theta(w^t, u) \geq 0 \quad \forall u \in D_\varepsilon(w^t). \tag{4.8}$$

Proof. For contradiction, suppose that for each j, there is a point $u^j \in D_\varepsilon(w^j)$ such that

$$\langle w^j - v, u^j - 0.5(w^j + v) \rangle < 0. \tag{4.9}$$

Since D is bounded, so is $\{u^j\}$. Hence $\{u^j\}$ has limit points and, without loss of generality, we suppose that $\{u^j\} \rightarrow \tilde{u}$. Then, $\tilde{u} \in D$ since D is closed. Next, by definition, it holds that

$$h_i(w^j) + \langle \nabla h_i(w^j), u^j - w^j \rangle \leq 0 \quad \forall i \in I_\varepsilon(w^j).$$

Since $\{w^j\} \rightarrow p^*$, we have $I_0(p^*) \subseteq I_\varepsilon(w^j)$ for j large enough. It follows that

$$h_i(w^j) + \langle \nabla h_i(w^j), u^j - w^j \rangle \leq 0 \quad \forall i \in I_0(p^*)$$

for j large enough. The sequences $\{u^j\}$ and $\{w^j\}$ are bounded, hence, taking the limit $j \rightarrow \infty$ in the above inequality yields

$$h_i(p^*) + \langle \nabla h_i(p^*), \tilde{u} - p^* \rangle \leq 0 \quad \forall i \in I_0(p^*),$$

i.e., $\tilde{u} \in U_0(p^*)$. Analogously, taking the limit $j \rightarrow \infty$ in (4.9) yields $\theta(p^*, \tilde{u}) \leq 0$, which now contradicts (4.6). $\square$

Now, noting that $U \subseteq D_\varepsilon(w^j)$ for all w^j and $\varepsilon \geq 0$ and using Lemma 4.1.2 (i), we obtain the following immediately.

Corollary 4.1.1. *Suppose that all the assumptions of Proposition 4.1.2 hold. Then, for each $\varepsilon > 0$, there exists a number t such that (4.2) holds with $p = w^t$.*

Thus, in order to implement a feasible quasi-nonexpansive mapping with respect to U, it suffices to apply any descent iterative algorithm for approximating a unique solution of the problem (4.7). There are a number of rather rapidly convergent iterative algorithms for this problem, such as feasible direction methods, center methods, etc; see [225, 175, 28] for example. Given a starting point $w^0 \in U$, these algorithms generate a feasible sequence $\{w^j\}$, such that the values of the cost function decrease monotonically. Therefore, each of these algorithms can be taken as a basis for constructing feasible quasi-nonexpansive mappings. In fact, Proposition 4.1.2 provides the condition (4.8) under which such an algorithm will terminate with a point $w^t = p \in U$ satisfying (4.2). The problem of verifying (4.8) is rather easy, namely, it is equivalent to a linear programming problem.

To illustrate the approach, we now describe a finite procedure for finding a point $p \in U$ satisfying (4.2), which is based on the feasible direction method from [225, Section 7.5].

For any points $w, h \in R^n$, we set

$$\lambda(w, h) = \max\{\alpha \mid \alpha \geq 0, w + \alpha h \in U\}.$$

Procedure 4.3. *Data:* A point $v \notin U$ and a point $u \in U$.
Output: A point p.
Parameters: Numbers $\varepsilon > 0, \delta > 0, \beta \in (0,1), \mu_i > 0$ for $i = 1, \ldots, m$ and a polyhedron $D \supseteq U$.
Step 0(Initialization): Set $w^0 := u, \delta_0 = \delta, j := 0$.
Step 1 (Verification): Find the solution z^j of the problem

$$\min \rightarrow \langle w^j - v, z \rangle$$

subject to

$$h_i(w^j) + \langle \nabla h_i(w^j), z - w^j \rangle \leq 0 \qquad i \in I_\varepsilon(w^j),$$
$$z \in D.$$

If $\theta(w^j, z^j) \geq 0$, set $p := w^j$ and stop.
Step 2 (Direction finding): Find the solution (τ_j, h^j) of the problem:

$$\min \rightarrow \tau$$

subject to

$$\langle w^j - v, h \rangle \leq \tau,$$
$$\langle \nabla h_i(w^j), h \rangle \leq \mu_i \tau \quad i \in I_{\delta_j}(w^j),$$
$$|h_i| \leq 1 \quad i = 1, \ldots, n.$$

Step 3 (Null step): If $\tau_j \geq -\delta_j$, set $\delta_{j+1} := \beta\delta_j, k := k+1$ and go to Step 2.

Step 4 (Descent step): Set $\lambda_j := \min\{\lambda'_j, \lambda''_j\}$, where

$$\lambda'_j = \langle h^j, v - w^j \rangle / \|h^j\|^2, \lambda''_j = \lambda(w^j, h^j). \tag{4.10}$$

Set $w^{j+1} := w^j + \lambda_j h^j, \delta_{j+1} := \delta_j, k := k+1$ and go to Step 1.

Thus, both the auxiliary subproblems in Procedure 4.3 are obviously linear programming problems. Note that λ'_j in (4.10) solves the problem

$$\min_{\alpha \geq 0} \rightarrow \|v - (w^j + \alpha h^j)\|^2,$$

whereas the value of λ''_j can be found by one of the well-known one - dimensional minimization algorithms with any prescribed accuracy. Combining Proposition 4.1.2 and the convergence theorem for the feasible direction method (see [225, Section 7.6]), we obtain the following finite termination property for Procedure 4.3.

Proposition 4.1.3. *Suppose that the set U is bounded. Then, Procedure 4.3 is finite and its final point $p = w^t \in U$ satisfies (4.2).*

4.2 Error Bounds for Linearly Constrained Problems

In Proposition 1.3.2, some classes of VI's on polyhedral sets satisfying Assumption (A5) were indicated. Assumption (A5) provides a linear rate of convergence for most methods of Chapter 1. We now intend to consider this assumption in detail and give the proof of Proposition 1.3.2.

4.2.1 Preliminaries

For the convenience of the reader, we recall the formulations of Assumptions (A1) – (A5) of Sections 1.2 and 1.3.

We consider VI (1.1), where U and V are nonempty, convex and closed subsets of R^n, $U \subseteq V$, $G : V \to R^n$ is a locally Lipschitz continuous mapping. Also, we assume that $U^* \neq \emptyset$. Next, we have a family of mappings $\{T_k : V \times V \to R^n\}$ satisfying the following assumptions for each $k = 0, 1, \ldots$:

   (A1) $T_k(u, \cdot)$ *is strongly monotone with constant $\tau'_k > 0$ for every $u \in U$;*
   (A2) $T_k(u, \cdot)$ *is Lipschitz continuous with constant $\tau''_k < \infty$ for every $u \in U$;*
   (A3) $T_k(u, u) = 0$ *for every $u \in U$;*
   (A4) $0 < \tau' \leq \tau'_k \leq \tau''_k < \tau'' < \infty$.
We now consider the following assumption.

   (A5) *Let $\{T_k\}$ be a sequence of mappings satisfying Assumptions (A1) – (A4) and let $z_\lambda(u)$ denote a (unique) solution to the problem*

$$\langle G(u) + \lambda^{-1} T_k(u, z_\lambda(u)), v - z_\lambda(u) \rangle \geq 0 \quad \forall v \in U, \tag{4.11}$$

with $0 < \lambda' \leq \lambda \leq \lambda'' < \infty$. Then there exist a number $\delta > 0$ and a number $\mu > 0$, which can only depend on δ, λ', and λ'', such that for any point $u \in V \supseteq U$, the inequality

$$\langle T_k(u, z_\lambda(u)), z_\lambda(u) - u \rangle \geq \mu \, d(u, U^*)^2$$

holds whenever $\langle T_k(u, z_\lambda(u)), z_\lambda(u) - u \rangle \leq \delta$.

   For brevity, we set

$$
\begin{aligned}
W_\lambda(u) &= \langle T_k(u, z_\lambda(u)), z_\lambda(u) - u \rangle, \\
\tilde{W}_\lambda(u) &= \|z_\lambda(u) - u\|^2.
\end{aligned}
$$

According to (A1) – (A4), we have

$$\tau' \|z_\lambda(u) - u\|^2 \leq \langle T_k(u, z_\lambda(u)), z_\lambda(u) - u \rangle \leq \tau'' \|z_\lambda(u) - u\|^2 \quad \forall u \in V,$$

or equivalently,

$$\tau' \tilde{W}_\lambda(u) \leq W_\lambda(u) \leq \tau'' \tilde{W}_\lambda(u) \quad \forall u \in V. \tag{4.12}$$

Due to (4.12), we can replace $W_\lambda(u)$ in (A5) with $\tilde{W}_\lambda(u)$. Moreover, it suffices to consider Assumption (A5) only with $\lambda = 1$ in (4.11). More precisely, if (A5) holds for $\lambda' = \lambda'' = 1$, then it holds for any fixed λ' and λ'' such that $0 < \lambda' \leq \lambda'' < \infty$, as the following lemma states.

Lemma 4.2.1. *There exist constants $\beta' > 0$ and $\beta'' < \infty$, depending on λ', λ'' and τ', τ'', such that*

$$\beta'\tilde{W}_1(u) \leq \tilde{W}_\lambda(u) \leq \beta''\tilde{W}_1(u). \tag{4.13}$$

Proof. By definition,

$$\langle G(u) + \lambda^{-1}T_k(u, z_\lambda(u)), z_1(u) - z_\lambda(u) \rangle \geq 0$$

and

$$\langle G(u) + T_k(u, z_1(u)), z_\lambda(u) - z_1(u) \rangle \geq 0.$$

Adding these inequalities gives

$$\langle \lambda^{-1}T_k(u, z_\lambda(u)) - T_k(u, z_1(u)), z_1(u) - z_\lambda(u) \rangle \geq 0. \tag{4.14}$$

On the one hand, it follows that

$$\begin{aligned}
0 \;\leq\; & \langle T_k(u, z_\lambda(u)) - T_k(u, z_1(u)), z_1(u) - z_\lambda(u) \rangle \\
+ \; & (1-\lambda)\langle T_k(u, z_1(u)) - T_k(u, u), z_1(u) - z_\lambda(u) \rangle \\
\leq\; & -\tau'\|z_\lambda(u) - z_1(u)\|^2 + |1 + \lambda''|\tau''\|z_1(u) - u\|\|z_\lambda(u) - z_1(u)\|,
\end{aligned}$$

hence

$$((1 + \lambda'')\tau''/\tau')\|z_1(u) - u\| \geq \|z_\lambda(u) - z_1(u)\| \geq \|z_\lambda(u) - u\| - \|z_1(u) - u\|.$$

Therefore, the right inequality in (4.13) holds with $\beta'' = (1 + (1 + \lambda'')\tau''/\tau')$. On the other hand, (4.14) also yields

$$\begin{aligned}
0 \;\leq\; & \langle T_k(u, z_1(u)) - T_k(u, z_\lambda(u)), z_\lambda(u) - z_1(u) \rangle \\
+ \; & \frac{\lambda - 1}{\lambda}\langle T_k(u, z_\lambda(u)) - T_k(u, u), z_\lambda(u) - z_1(u) \rangle \\
\leq\; & -\tau'\|z_1(u) - z_\lambda(u)\|^2 + [(1 + \lambda'')\tau''/\lambda']\|z_\lambda(u) - u\|\|z_1(u) - z_\lambda(u)\|.
\end{aligned}$$

It follows that

$$\begin{aligned}
[(1 + \lambda'')\tau''/(\lambda'\tau')]\,\|z_\lambda(u) - u\| \;&\geq\; \|z_1(u) - z_\lambda(u)\| \\
&\geq\; \|z_1(u) - u\| - \|z_\lambda(u) - u\|.
\end{aligned}$$

Therefore, the left inequality in (4.13) also holds with $\beta' = 1/(1 + [(1 + \lambda'')\tau''/(\lambda'\tau')])$. □

Combining (4.12) and (4.13), we conclude that it suffices to obtain Assumption (A5) for $W_1(u)$ or $\tilde{W}_1(u)$ and (A5) will hold for $W_\lambda(u)$ with $\lambda \in [\lambda', \lambda'']$.

4.2.2 Error Bounds

We first give a reduced version of Assumption (A5) in the case where $T_k(u, z) = z - u$ and $\lambda = 1$. Clearly, due to Lemma 1.2.2, $z_1(u)$ is then the projection of the point $u - G(u)$ onto U. Namely, let us consider the following property.

(A5') *Let u be a point of V. There exist a number $\delta' > 0$ and a number $\mu' > 0$, which can depend on δ, such that the inequality*

$$\|u - \pi_U(u - G(u))\| \geq \mu' d(u, U^*)$$

holds whenever $\|u - \pi_U(u - G(u))\| \leq \delta'$.

This projective error bound was investigated by many authors. We now give the conditions which provide for (A5') to hold.

Proposition 4.2.1. *[211, Theorem 2] Suppose that U is a polyhedral set. Suppose also that at least one of the following assumptions holds:*
  (i) G is affine;
  (ii) $G(u) = E^T F(Eu) + q$, where E is an $m \times n$ matrix with no zero column, q is a vector in R^n and $F : R^m \to R^m$ is a strongly monotone Lipschitz continuous.
  Then Assumption (A5') is fulfilled.

Thus, Proposition 1.3.2 is an extension of the above results. Hence, to prove Proposition 1.3.2 it suffices to show that $\tilde{W}_1(u)$ is a majorant for $\|u - \pi_U(u - G(u))\|^2$.

Lemma 4.2.2. *Suppose that a sequence of mappings $T_k : V \times V \to R^n$ satisfies (A1) – (A4). Then*

$$(2 + \tau'') \|u - z_1(u)\| \geq \|u - \pi_U(u - G(u))\|. \tag{4.15}$$

Proof. For brevity, set $w = \pi_U(u - G(u))$. By definition, we have

$$\langle G(u) + T_k(u, z_1(u)), w - z_1(u) \rangle \geq 0$$

and

$$\langle G(u) + w - u, z_1(u) - w \rangle \geq 0.$$

Adding these inequalities gives

$$\langle w - z_1(u) + z_1(u) - u - T_k(u, z_1(u)), z_1(u) - w \rangle \geq 0,$$

or equivalently,

$$\langle z_1(u) - u, z_1(u) - w \rangle + \langle T_k(u, u) - T_k(u, z_1(u)), z_1(u) - w \rangle \geq \|z_1(u) - w\|^2.$$

It follows that

$$(1 + \tau'') \, \|z_1(u) - u\| \, \|z_1(u) - w\| \geq \|z_1(u) - w\|^2 \,,$$

hence

$$(1 + \tau'') \, \|z_1(u) - u\| \geq \|w - u\| - \|u - z_1(u)\| \,,$$

and the result follows.   □

We are now in a position to establish the result of Proposition 1.3.2. For the sake of convenience, we give its formulation here again.

Proposition 4.2.2. *Suppose that U is a polyhedral set. Suppose also that at least one of the following assumptions holds:*
  (i) G is affine;
  (ii) $G(u) = E^T F(Eu) + q$, where E is an $m \times n$ matrix with no zero column, q is a vector in R^n and $F : R^m \to R^m$ is a strongly monotone Lipschitz continuous.
  Then Assumption (A5) is fulfilled.

Proof. Due to (4.12) and Lemma 4.2.1 it suffices to show that (A5) is true for $\tilde{W}_1(u)$ instead of $W_\lambda(u)$. Let μ' and δ' be constants associated to (A5′). Set $\delta = [\delta'/(2 + \tau'')]^2$. If $\tilde{W}_1(u) \leq \delta$, then

$$\|u - \pi_U(u - G(u))\| \leq (2 + \tau'') \, \|u - z_1(u)\| \leq \delta'$$

due to (4.15). From Proposition 4.2.1 it now follows that

$$\|u - \pi_U(u - G(u))\| \geq \mu' d(u, U^*).$$

Combining this inequality with (4.15) gives

$$\|u - z_1(u)\| \geq [\mu'/(2 + \tau'')] \, d(u, U^*),$$

i.e., Assumption (A5) is fulfilled.   □

In the case where $U^* = U^d$, there exists a projection of every point onto U^* since it becomes convex and closed due to Proposition 1.1.2. Then we can replace the distance $d(u, U^*)$ with its explicit form $\|u - \pi_{U^*}(u)\|$.

4.3 A Relaxation Subgradient Method Without Linesearch

In this section, we describe the simplest relaxation subgradient method for convex minimization, which was proposed in [95, 97]. Although most investigations in nonsmooth optimization are devoted to the bundle methods (e.g., see [91, 78]), this method also has own certain advantages. In particular, it turns out to be very suitable for constructing auxiliary procedures in CR methods for GVI's. We establish its convergence properties and a complexity estimate.

4.3.1 Description of the Method

Let $f : R^n \to R$ and $h : R^n \to R$ be convex, but not necessarily differentiable functions. We consider the minimization problem

$$\min \to \{f(x) \mid x \in D\} \tag{4.16}$$

where

$$D = \{x \in R^n \mid h(x) \leq 0\}. \tag{4.17}$$

In addition, we suppose that there exists a point $\bar{x} \in X$ such that $h(\bar{x}) < 0$ and that the set

$$D(\bar{x}) = \{y \in D \mid f(y) \leq f(\bar{x})\}$$

is bounded. It follows that the convex minimization problem (4.16), (4.17) is solvable and that

$$f^* = \inf \{f(x) \mid x \in D\} > -\infty.$$

Next, for any $x \in R^n$ we define the set

$$Q(u) = \begin{cases} \partial f(x) & \text{if } h(x) \leq 0, \\ \partial h(x) & \text{if } h(x) > 0. \end{cases} \tag{4.18}$$

Then, the relaxation subgradient method for solving problem (4.16), (4.17) can be described as follows.

Method 4.1. *Step 0 (Initialization)* : Choose a point $x^0 \in D$, a number $\theta \in (0, 1)$, positive bounded sequences $\{\varepsilon_l\}, \{\eta_l\}$. Set $y^0 := x^0$, $k := 0$, $\lambda(k) := 1$.
  Step 1 (Auxiliary procedure) :
    Step 1.0 : Set $i := 0$, $w^0 := x^k$, $l := \lambda(k)$. Choose $q^0 \in Q(w^0)$, set $p^0 := q^0$.
    Step 1.1 : If $\|p^i\| \leq \eta_l$, set $x^{k+1} := y^l := x^k$, $\lambda(k) := l + 1$, $k := k + 1$ and go to Step 1 (*null step*).
    Step 1.2 : Set $w^{i+1} := w^0 - \varepsilon_l p^i / \|p^i\|$. If

$$f(w^{i+1}) < f(w^0) - \theta \varepsilon_l \|p^i\|, h(w^{i+1}) \leq 0, \tag{4.19}$$

then set $\lambda(k+1) := l$ and go to Step 2 (*descent step*).

  Step 1.3 : Choose $q^{i+1} \in Q(w^{i+1})$, set

$$p^{i+1} := \text{Nr conv}\{p^i, q^{i+1}\},$$

$i := i + 1$ and go to Step 1.1.

  Step 2 (Main iteration) : Set $x^{k+1} := w^{i+1}$, $k := k+1$ and go to Step 1.

Note that the structure of Method 4.1 is similar to that of Methods 2.2 – 2.4. The main difference is in the main iteration. Moreover, Method 4.1 also includes no line search and requires only two vectors for computing a new direction. We will call one change of the index i an inner step. Also, we denote by k_l an iteration number such that $\lambda(k_l) = l$, $\lambda(k_l + 1) = l + 1$.

4.3.2 Convergence

We first establish two auxiliary properties of vector sequences.

Lemma 4.3.1. *Let $\{p^i\}$ and $\{q^i\}$ be sequences in R^n such that*

$$\|q^i\| \leq C < \infty, p^{i+1} = \text{Nr conv}\{p^i, q^{i+1}\}, \quad i = 0, 1, \ldots$$
$$p^0 = q^0. \tag{4.20}$$

Then

$$p^i \in \text{conv}\{q^0, \ldots, q^i\}, \quad i = 0, 1, \ldots$$

Proof. For brevity, we set $Q_i := \text{conv}\{q^0, \ldots, q^i\}$. We prove the assertion by induction on i. It is clear that $p^0 \in Q_0$. By definition, $q^i \in Q_i$ and $Q_{i-1} \subseteq Q_i$. If we suppose that $p^{i-1} \in Q_{i-1}$, then $p^{i-1} \in Q_i$ and $p^i \in \text{conv}\{p^{i-1}, q^i\} \subseteq Q_i$, as desired. □

Lemma 4.3.2. *Let $\{p^i\}$ and $\{q^i\}$ be sequences in R^n such that (4.20) holds and*

$$\langle q^{i+1}, p^i \rangle \leq \theta \|p^i\|^2, \quad i = 0, 1, \ldots, \ \theta \in (0, 1). \tag{4.21}$$

Then

$$\|p^i\| \leq C/((1-\theta)\sqrt{i+1}) \quad \text{for } i = 0, 1, \ldots \tag{4.22}$$

Proof. We prove inequality (4.21) by induction on i. It is clear that $\|p^0\| = \|q^0\| \leq C/(1-\theta)$. Next, by (4.20), (4.21), we have

$$\|p^i\|^2 = \min_{\mu \in [0,1]} \|\mu p^{i-1} + (1-\mu)q^i\|^2$$

$$= \min_{\mu \in [0,1]} (\mu^2 \|p^{i-1}\|^2 + 2\mu(1-\mu)\langle p^{i-1}, q^i \rangle + (1-\mu)^2 \|q^i\|^2)$$

$$\leq \min_{\mu \in [0,1]} \varphi(\mu),$$

where

$$\varphi(\mu) = \mu^2 \|p^{i-1}\|^2 + 2\theta\mu(1-\mu)\|p^{i-1}\|^2 + (1-\mu)^2 C^2.$$

From Lemma 4.3.1 and (4.20) it follows that $\|p^{i-1}\| \leq C$, hence

$$\varphi''(\mu) = 2((1-\theta)\|p^{i-1}\|^2 + C^2 - \theta\|p^{i-1}\|^2) > 0.$$

So, φ is strictly convex and has a global minimum at

$$\mu^* = (1 + (1-\theta)\|p^{i-1}\|^2/(C^2 - \theta\|p^{i-1}\|^2))^{-1}.$$

Since $\mu^* \in [0, 1]$, we obtain

$$
\begin{aligned}
\|p^i\|^2 &\leq \varphi(\mu^*) \\
&= C^2 - (C^2 - \theta\|p^{i-1}\|^2)^2/((1-\theta)\|p^{i-1}\|^2 + C^2 - \theta\|p^{i-1}\|^2) \\
&= \|p^{i-1}\|^2(C^2 - \theta^2\|p^{i-1}\|^2)/((1-\theta)\|p^{i-1}\|^2 + C^2 - \theta\|p^{i-1}\|^2) \\
&= \|p^{i-1}\|^2(1 + (1-\theta)^2\|p^{i-1}\|^2/(C^2 - \theta^2\|p^{i-1}\|^2))^{-1} \\
&\leq \|p^{i-1}\|^2(1 + (1-\theta)^2\|p^{i-1}\|^2/C^2)^{-1} \\
&= (\|p^{i-1}\|^{-2} + (1-\theta)^2/C^2)^{-1}.
\end{aligned}
$$

If we now suppose that $\|p^{i-1}\| \leq C/((1-\theta)\sqrt{i})$, then, by the above inequality, we must have

$$\|p^i\|^2 \leq ((i+1)(1-\theta)^2/C^2)^{-1}.$$

Thus (4.22) holds too. $\square$

We now obtain accuracy measures for Method 4.1.

Lemma 4.3.3. *For each $l = 1, 2, \ldots$, there exists a number $\tau_l \in [0, 1]$ such that the following inequalities hold:*

$$\tau_l(f(y^l) - f^*) \leq \eta_l\|y^l - x^*\| + (1+\theta)C_l\varepsilon_l, \tag{4.23}$$

and

$$\tau_l(f(\bar{x}) - f^* - h(\bar{x})) \geq -h(\bar{x}) - \eta_l\|\bar{x} - y^l\| - (1+\theta)\varepsilon_l C_l, \tag{4.24}$$

where x^* is any solution to problem (4.16), (4.17),

$$C_l = \sup\left\{ \|g\| \mid g \in \text{conv} \bigcup_{\|x-y^l\|\leq\varepsilon_l} Q(x) \right\}. \tag{4.25}$$

Proof. Let us consider the k_lth iteration of the method. Then $x^{k_l} = y^l$ and by Lemma 4.3.1, for each $t = 0, 1, \ldots$,

$$p^t = \sum_{i=0}^{t} \beta_i q^i, \quad \sum_{i=0}^{t} \beta_i = 1; \beta_i \geq 0 \quad \text{for} \quad i = 0, \ldots, t. \tag{4.26}$$

By definition, at the k_lth iteration, the auxiliary procedure produces a null step, i.e., there exists a number $\tilde{t}$ such that

$$\|p^{\tilde{t}}\| \leq \eta_l. \tag{4.27}$$

Define the index sets

$$I = \{0, \ldots, \tilde{t}\}, \bar{I} = \{i \in I \mid w^i \in D\}, \tilde{I} = \{i \in I \mid w^i \notin D\}.$$

Then, by (4.18), we have

$$q^i \in \begin{cases} \partial f(w^i) & \text{if } i \in \bar{I}, \\ \partial h(w^i) & \text{if } i \in \tilde{I}. \end{cases} \tag{4.28}$$

Set

$$\tau_l = \sum_{i \in \bar{I}} \beta_i. \tag{4.29}$$

It is obvious that $\tau_l \in [0, 1]$. Observe also that

$$\|w^i - y^l\| \leq \varepsilon_l \quad \text{for} \quad i \in I. \tag{4.30}$$

Fix $y \in D$. In case $i \in \bar{I}$, (4.19) does not hold, hence, using (4.25), (4.28) and (4.30) gives

$$\begin{aligned}
f(y) - f(y^l) &\geq \langle q^i, y - w^i \rangle + f(w^i) - f(y^l) \\
&\geq \langle q^i, y - y^l \rangle + \langle q^i, y^l - w^i \rangle - \theta \varepsilon_l \|p^{i-1}\| \\
&\geq \langle q^i, y - y^l \rangle - (\|q^i\| + \theta \|p^{i-1}\|)\varepsilon_l \\
&\geq \langle q^i, y - y^l \rangle - (1 + \theta)C_l \varepsilon_l.
\end{aligned}$$

Analogously, in case $i \in \tilde{I}$, using (4.25), (4.28) and (4.30) gives

$$\begin{aligned}
0 &\geq h(y) - h(w^i) \geq \langle q^i, y - w^i \rangle \geq \langle q^i, y - y^l \rangle - \varepsilon_l \|q_i\| \\
&\geq \langle q^i, y - y^l \rangle - \varepsilon_l C_l.
\end{aligned}$$

Setting $y = x^*$ and summing these inequalities, multiplied by β_i, over $i \in I$ and taking into account (4.26), (4.27) and (4.29), we have

$$\tau_l(f(x^*) - f(y^l)) \geq \langle p^{\tilde{t}}, x^* - y^l \rangle - (1 + \theta)\varepsilon_l C_l \geq -\eta_l \|x^* - y^l\| - (1 + \theta)\varepsilon_l C_l,$$

i.e. (4.23) holds. Analogously, setting $y = \bar{x}$ and summing these inequalities, multiplied by β_i, over $i \in I$ and taking into account (4.26), (4.27) and (4.29), we have

$$\tau_l(f(\bar{x}) - f(y^l)) + (1 - \tau_l)h(\bar{x}) \geq -\eta_l \|\bar{x} - y^l\| - (1 + \theta)\varepsilon_l C_l,$$

i.e. (4.24) holds, too. $\qquad \square$

Lemma 4.3.4. *Let k and l be fixed. Then the number of inner steps does not exceed the value*

$$C^2_{k,l}/(\eta_l(1-\theta))^2, \tag{4.31}$$

where

$$C_{k,l} = \sup\left\{\|g\| \mid g \in \operatorname{conv} \bigcup_{\|v-x^k\|\le\varepsilon_l} Q(v)\right\}.$$

Proof. Let (4.19) do not hold. If $h(w^{i+1}) \le 0$, then

$$
\begin{aligned}
\theta\varepsilon_l\|p^i\| &\ge f(w^0) - f(w^{i+1}) \ge \langle q^{i+1}, w^0 - w^{i+1}\rangle\\
&= \varepsilon_l\langle q^{i+1}, p^i/\|p^i\|\rangle,
\end{aligned}
$$

which gives (4.21). If $h(w^{i+1}) > 0$, then

$$0 \ge h(w^0 - h(w^{i+1}) \ge \langle q^{i+1}, w^0 - w^{i+1}\rangle = \varepsilon_l\langle q^{i+1}, p^i/\|p^i\|\rangle,$$

which also gives (4.21). Thus, $\{q^i\}$ and $\{p^i\}$ satisfy the assumptions of Lemma 4.3.2, so that the desired inequality (4.31) now follows from (4.22) and the stopping test in Step 1.1. $\qquad\square$

We are now ready to establish a convergence result for Method 4.1.

Theorem 4.3.1. *Let $\{\varepsilon_l\}$ and $\{\eta_l\}$ be chosen by*

$$\{\varepsilon_l\} \searrow 0, \{\eta_l\} \searrow 0.$$

Then:
   (i) For each $l = 1, 2, \ldots$, the number of inner steps is finite.
   (ii) It holds that

$$\lim_{l\to\infty} f(y^l) = f^*.$$

Proof. On account of Lemma 4.3.4, the number of inner steps is finite when k and l are fixed. On the other hand, given a fixed l, the number of changes of the index k does not exceed the value

$$(f(y^{l-1}) - f^*)/(\theta\varepsilon_l\eta_l)$$

because of (4.19). This proves part (i). Next, since $D(\bar{x})$ is bounded, so is $D(y^0)$ (e.g., see [68, Chapter 1, Lemma 2.1]). Therefore, $\{y^l\}$ is infinite and bounded. Taking the lower limit $l \to \infty$ in (4.24) now gives

$$\liminf_{l\to\infty} \eta_l \ge \tau' = -h(\bar{x})/(f(\bar{x}) - f^* - h(\bar{x})) > 0.$$

Taking the limit $l \to \infty$ in (4.23) and using the above inequality then gives

$$\lim_{l\to\infty} f(y^l) = f^*,$$

i.e. part (ii) is also true. The proof is complete. $\qquad\square$

4.3.3 Rate of Convergence

By construction, Method 4.1 requires no linear searches and no a priori constants to be evaluated. Nevertheless, we show that such a method attains a linear rate of convergence. More precisely, we give a complexity estimate which corresponds to such a rate. Given a starting point x^0 and a number $\delta > 0$, we define the complexity of the method, denoted by $N(\delta)$, as the total number of inner steps at the $l(\delta)$ major iterations, i.e. when

$$(f(y^l) - f^*)/(f(y^0) - f^*) \leq \delta \quad \text{for} \quad l > l(\delta).$$

Theorem 4.3.2. *Suppose that $D = R^n$, the function f is strongly convex with constant κ, its gradient is Lipschitz continuous with constant L. Let the parameters of Method 4.1 be chosen by*

$$\varepsilon_l = \nu^l \varepsilon', \eta_l = \nu^l \eta', l = 0, 1, \ldots; \varepsilon' > 0, \eta' > 0, \nu \in (0,1). \tag{4.32}$$

Then

$$N(\delta) \leq B_1 \ln(1/\delta)/(\nu^2 \ln(\nu^{-2})), \tag{4.33}$$

where B_1 is independent of δ and ν.

Proof. By using the same argument as that in the proof of Theorem 4.3.1, we see that $\{y^l\}$ is infinite and bounded. Besides, under the assumptions of the present theorem, problem (4.16) has a unique solution due to Corollary 1.1.1. Let k and l be fixed. Then,

$$\|q^i\| \leq \|\nabla f(x^k)\| + L\varepsilon_l, \|\nabla f(x^k)\| \leq \|p^t\| + L\varepsilon_l \tag{4.34}$$

for $i = 0, 1, \ldots, t$. It follows that

$$\|q^i\| \leq \|p^t\| + 2L\varepsilon_l$$

for $i = 0, 1, \ldots, t$. Hence, if $\|p^t\| \geq \eta_l$, then we have

$$C_{k,l}/\|p^t\| \leq (\|p^t\| + 2L\varepsilon_l)/\|p^t\| \leq 1 + 2L\varepsilon'/\eta'. \tag{4.35}$$

Following the proof of Lemma 4.3.4 we now conclude that the number of inner steps does not exceed the value

$$[(1 + 2L\varepsilon')/(\eta(1 - \theta))]^2 + 1. \tag{4.36}$$

Next, by definition $\eta_l = 1$, hence (4.23) now yields

$$f(y^l) - f^* \leq \eta_l \|y^l - x^*\| + (1 + \theta)C_l \varepsilon_l. \tag{4.37}$$

But ∇f is now strongly monotone with constant κ due to Proposition 1.1.9, hence, for each $x \in R^n$,

$$\kappa\|x - x^*\|^2 \le \langle \nabla f(x) - \nabla f(x^*), x - x^* \rangle \le \|\nabla f(x)\|\|x - x^*\|,$$

i.e.,

$$\kappa\|x - x^*\| \le \|\nabla f(x)\|.$$

Combining this inequality with (4.32), (4.34), (4.35), and (4.37) gives

$$
\begin{aligned}
f(y^l) - f^* &\le \eta_l \|\nabla f(y^l)\|/\kappa + (1+\theta)C_l\varepsilon_l \\
&\le \eta_l(\eta_l + L\varepsilon_l)/\kappa + (1+\theta)(\eta_l + 2L\varepsilon_l)\varepsilon_l \\
&\le B_2\nu^{2l}
\end{aligned}
$$

where $B_2 = \eta'(\eta' + L\varepsilon')/\kappa + (1+\theta)(\eta' + 2L\varepsilon')\varepsilon'$. Hence,

$$k_{l+1} - k_l \le (f(y_l) - f^*)/(\theta\varepsilon_{l+1}\eta_{l+1}) \le B_2/\theta\nu^2$$

and, taking into account (4.36), we conclude that the number of inner steps for each fixed l does not exceed the value

$$B_1/\nu^2,$$

where $B_1 = B_2\left[(1 + 2L\varepsilon')(\eta'(1-\theta))^2 + 1\right]$.

Therefore, the total amount of inner steps at the $l(\delta)$ "big" iterations is the following:

$$k(\delta)B_1/\nu^2 \le B_1 \ln(1/\delta)/(\nu^2 \ln(\nu^{-2})).$$

Hence, the upper bound (4.33) hold and the proof is complete.   □

Thus, the method attains a complexity estimate which is equivalent to linear convergence with respect to inner steps.

Bibliographical Notes

Due to the diversity and a great number of works, these notes are not naturally full. When considering standard results, we refer to the well-known books.

Chapter 1.

1.1. Most of the results of this section are rather standard, see [70, 55].

1.1.1. The variational inequality problem was first considered by G. Fichera [56] and G. Stampacchia [204]. Various existence and uniqueness results for VI's can be for instance found in [6, 9, 33, 49, 68, 89, 213]. Lemma 1.1.1 and Example 1.1.2 were obtained in [121]. Figure 1.1 was also taken from [121].

1.1.2. Various classes of complementarity problems are considered in [34, 80]. The fixed point problem is one of the main problems in Nonlinear Analysis; its theory, methods and applications can be for example found in [6, 9, 15, 159, 161, 164, 210, 215]. The result of Proposition 1.1.8 contained, for example, in [21]. For a review of differentiable optimization theory and methods see, for example, [16, 28, 64, 65, 158, 165, 175, 179, 183, 214, 220, 222, 225]. The result of Theorem 1.1.1 is contained in [147, p.118]. Example 1.1.4 is well known, e.g., see [68, p.262]; Proposition 1.1.10 is known as the Kuhn-Tucker saddle point theorem; its variants can be for example found in [214, 16]. There exist a number of variants of Proposition 1.1.11, see for example [70, Proposition 2.2].

1.2. In this section, we follows [113, 115] in general.

1.2.1. The Newton method for nonlinear equations is described in [84]; see also [165]. Properties of various Newton-like algorithms for variational inequalities are summarized in [170]. The result of Lemma 1.2.2 is well known, e.g., see [59]. Various descent schemes for strongly monotone variational inequalities were considered by several authors; e.g., see [32, 36, 59, 173, 224].

1.2.2. Algorithm (1.33) is due to I.I. Eremin [50] and B.T. Polyak [178]. There are a number of algorithms for linear equations, linear and convex inequalities, which generate iteration sequences with monotone decrease of the Euclidean distance to each solution; see [83, 30, 2, 152, 50]. These algorithms are also called Fejer-monotone ones, they were investigated in detail in [51, 196, 92, 13]. Nevertheless, all these methods are one-level ones in the sense that they use no auxiliary procedures, hence, they cannot be directly applied to variational inequalities. The combined relaxation approach for variational inequalities and the corresponding basic scheme following the rules (1.35), (1.36) (with $P_k(\cdot) = \pi_W(\cdot)$) were first proposed in [103]. The first method for solving nonlinear equations converging to a solution of just the dual variational inequality under the additional strong monotonicity (respectively, monotonicity) assumption was proposed by A.A. Abramov and A.V. Gaipova in [1] (respectively by A.S. Nemirovskii in [155]). N.Z. Shor in [200] first suggested to find a point u^* satisfying (1.2) with the ellipsoid method. The methods for solving variational inequalities, which converge to a solution of the dual problem (1.2) without additional monotonicity assumptions were first proposed in [103]. Various classes of non-expansive (but not necessarily feasible) mappings were investigated in detail in [215]. The results of Proposition 1.2.1 are well known, e.g., see [28, Chapter 1, Section 2.1].

1.3. The first implementable combined relaxation methods were proposed in [103]. The auxiliary procedures of these are based on an iteration of the projection method, Frank-Wolfe type method, and symmetric Newton method, respectively. The first corresponds to Method 1.1 with $T_k(u, z) = z - u$. Also, in [103], it was noticed that most descent methods in optimization can be served for constructing auxiliary procedures in combined relaxation methods, including various variants of the gradient projection methods. For instance, in the case where $T_k(u, z) = z - u$, the auxiliary procedure in Method 1.1 is a modification of Schemes 1 and 4 from [42, Chapter 3, Section 2], whereas the auxiliary procedure in Method 1.2 is a modification of Scheme 3 from [42, Chapter 3, Section 2].

1.3.1. We follows [115].

1.3.2. The results of this section are some modifications of those in [115]. In the case where $T_k(u, z) = z - u$, Assumption (A5) was considered by many authors and the results similar to Propositions 1.3.1 and 1.3.2 were derived for other methods; e.g., see [211, 169]. Theorems 1.3.2 and 1.3.3 generalize the corresponding assertions in [115].

1.3.3. In the case where $T_k(u, z) = z - u$, similar methods for (pseudo) monotone VI's were proposed in [81, 205]; see also [113].

1.3.4. The fact of accelerating convergence of a Fejer-type relaxation method for matrix games after incorporating the projections onto the manifold $\{x \mid \sum x_i = 1\}$ was noticed in [129]. In the variational inequality case, such techniques combining ideas of different projection methods (see [190, 136]) were developed in [205, 206, 207, 110, 116].

1.4. In this section, we follows [113, 116, 119, 120] in general. In [74]–[76], B. He proposed several methods, which require no line searches, for linear complementarity problems and linear variational inequalities. These methods exploit the same idea of Fejer-monotone approximation to all solutions and are close to the methods of Section 1.4, however, they do not seem to be extended to the nonlinear case.

1.4.1. In the case where $T_k(u, z) = z - u$, a similar method was proposed in [206, 207]. Method 1.3 is a modification of that in [119].

1.4.2. In the case where $T_k(u,z) = z - u$ and $V = U$ (respectively, $V = R^n$), similar methods were proposed in [207, 203, 77]. Method 1.4 is a modification of that in [120] and all the results of this subsection, with the exception of Theorem 1.4.5 (i), are taken from [120], with the corresponding little modifications. This material is included with permission from Gordon and Breach Publishers.

1.5. Method 1.5 is an extension of that in [103]. Frank-Wolfe type algorithms for equilibrium and variational inequality problems under additional assumptions were considered in [227].

1.6. An implementable CR method for a convex-concave equilibrium problem with smooth nonlinear constraints, which is based on an iteration of the feasible direction method and incorporates feasible non-expansive operators, was proposed in [114, 116].

Chapter 2.

2.1. Most of results of this section are rather standard; see [6, 9, 21, 31, 53, 70, 185]. Variational inequalities with multivalued monotone mappings were first investigated by F.E. Browder in [24].

2.1.1. Various existence and uniqueness results for GVI are contained, for example, in [21, 38, 53, 198, 219]. Lemma 2.1.1 was obtained in [121]. The relationships with vector variational inequalities were investigated in [127].

2.1.2. Various generalizations of the gradient mapping in the nonsmooth case were proposed by many authors, e.g., see [31, 41, 43, 44, 182, 185, 201]. Proposition 2.1.13 was obtained in [188, 10].

2.1.3. General equilibrium problems were investigated by many authors; e.g., see [9, 14, 18, 21, 23, 52, 126, 161, 164, 223, 226]. H. Nikaido and K. Isoda in [160] first proposed to convert the Nash equilibrium problem in a non-cooperative game into a general equilibrium problem with the help of (2.16) and (2.17). The mixed variational inequality problem (2.18) with affine F was originally considered by C. Lescarret in [135]. In [25], F.E. Browder first considered MVI's with nonlinear cost mappings. Afterwards, various classes of MVI's were studied by many authors; see [9, 32, 45, 49, 66, 162, 166]. The mapping G in (2.21) was originally introduced by J.B. Rosen [191] in the smooth case; its extensions were investigated by many authors; see [68, 137, 212]. The results which are close to some assertions of Proposition 2.1.17 are contained in [4, 94].

2.2. A general descent approach to constructing solution methods for MVI was considered in [173]. Some methods for solving MVI under additional strong monotonicity (convexity) assumptions were proposed in [163, 172], an auxiliary problem similar to (2.32) was considered in [172]. In [171], G.B. Passty suggested to combine splitting and averaging in order to obtain convergence without additional strong monotonicity assumptions. Note that Method 2.1 involves approximately the same computational expense per iteration but, unlike the averaging method, attains a linear rate of convergence. Assumption (B3) extends the concept of a sharp solution from [173, Proposition 7.28]. An extension of the sharp optimum condition for equilibrium problems was considered in [3]. A similar method for variational inequalities whose basic operator is the sum of a single- and a multi-valued operator was suggested in [125].

2.3. When obtaining the convergence properties of Method 2.1, we follows [106] and [118] in general.

2.4. The method of this section modifies and generalizes those in [106, 111, 122].

2.5. The method of this section modifies and generalizes that in [123]. The parts of Subsection 2.5.1 which are related to the statement of the problem and the definition of a pseudo P-monotone mapping and, also, all the results of Subsections 2.5.2 and 2.5.3 are taken from [123], with the corresponding modifications. This material is included with permission from Gordon and Breach Publishers.

Chapter 3.

3.1.1. The proximal mapping was introduced by J.J. Moreau in [150]. The proximal point method was proposed by B. Martinet [146] and further developed by R.T. Rockafellar [186, 187]. A general approach to constructing proximal point algorithms was proposed in [68, Chapter 5].

3.1.2. The iterative regularization method was proposed by A.B. Bakushinskii and B.T. Polyak in [12]. Comprehensive description of various regularization type methods is given in [11, 215]. The averaging method was proposed by R. Bruck in [26] and investigated by A.S. Nemirovskii in [155] and S.P. Uryas'yev in [212]. Some algorithms which can be viewed as intermediate ones between the iterative regularization method and the averaging method are described in [215]. A variant of the averaging method with feasible nonexpansive operators was proposed in [118].

3.1.3. The ellipsoid method was first proposed by D.B. Yudin and A.S. Nemirovskii [221] and by N.Z. Shor [199] for convex optimization problems. In [200], N.Z. Shor suggested this method to be applied to find a point u^* satisfying (2.2), in particular, to saddle point problems. For monotone GVI's, the ellipsoid method was developed by A.S. Nemirovskii in [155]. Descriptions of various center type methods for variational inequalities can be found in [143, 67, 133, 93]. Additional properties of the ellipsoid method are described in [201, 156, 197].

3.1.4. The extragradient method for VI was first proposed by G.M. Korpelevich [128] and developed by A.S. Antipin [3]. In the case of saddle point problems, the same idea of making use of extrapolation points to guarantee for the corresponding gradient-type method to converge was perhaps first proposed by T. Kose in [130]. A similar process was considered by K.J. Arrow and R.M. Solow in [5].

Recently, a number of descent methods for CP's and VI's were proposed; e.g. see [224, 60, 57, 19]. They use various merit functions for CP's and VI's. Some of these methods attain quadratic convergence. However, they are usually convergent under additional strong monotononicity or regularity assumptions. Rates of convergence of the descent methods [145, 19] adjusted for monotone VI's are now under examination.

3.2. Various economic equilibrium models were considered in [6, 37, 55, 70, 144, 154, 159, 177, 193, 209, 210].

3.2.1. Variants of the revealed preference condition (3.25) were considered in detail in [159, 177], for example.

3.2.2. In considering the equilibrium model with fixed budgets we follow [177, Chapters 2–4].

3.2.3. The linear exchange model is due to D. Gale [62, 63]. It was also considered in detail in [177, 181]. A fixed point type algorithm to find a solution of this model was proposed by B.C. Eaves in [48].

3.2.4. The general equilibrium model was proposed by H.E. Scarf (see [193]). Various applications and algorithms for this model are considered in [144].

3.2.5. The oligopolistic market equilibrium model originated by A. Cournot [35] for the case where $n = 2$ and all the functions are affine. Most solution methods for finding oligopolistic equilibria are based on reducing the initial problem to a parametric optimization problem, e.g., see [208, 153, 27].

3.3. Example 3.3.9 is an extension of Example 14 from [118]. Other results of numerical experiments with combined relaxation type methods and with the extragradient method are given in [74, 76, 88, 203, 205, 206, 207].

Chapter 4.

4.1. Procedures 4.1 and 4.2 were proposed in [131, 132], respectively. Another general approach to constructing feasible quasi-nonexpansive operators was considered in [124].

4.2. Several different approaches to deriving error bounds were proposed and studied in [184, 168, 140, 142, 120]. Most of them are based on the well-known Hoffman lemma from [79]. An excellent survey of various error bounds in optimization and related fields, including variational inequalities, is given in [169]. The results of this section can be viewed as extensions of those in [61, Lemma 1] for the simplest "projective" residual functions, see also [211].

4.3. Other variants of relaxation subgradient methods without linesearch (in the convex case) were proposed in [29, 90, 176].

4.2.1. We follow [95, 97]. The result similar to that of Lemma 4.3.2 in the case of $\theta \leq 0.5$ was obtained in [217].

4.2.2. The result of Theorem 4.3.2 was first obtained in [95]. Other complexity estimates for Method 4.1 were obtained in [95, 100, 105].

References

1. Abramov, A.A., Gaipova, A.N. (1972): On Solution of Some Equations Involving Discontinuous Monotone Transformations. Zhurnal Vychislitel'noi Matematiki i Matematicheskoi Fiziki 12, 204–207 (in Russian)
2. Agmon, S. (1954): The Relaxation Method for Linear Inequalities. Canadian Journal of Mathematics 6, 382–392
3. Antipin, A.S. (1995): On Convergence of Proximal Methods to Fixed Points of Extremal Mappings and Estimates of Their Rate of Convergence. Computational Mathematics and Mathematical Physics 35, 539–551
4. Antipin, A.S. (1997): Equilibrium Programming: Proximal Methods. Computational Mathematics and Mathematical Physics 37, 1285–1296
5. Arrow, K.J., Hurwicz, L., Uzava, H. (1958): Studies in Linear and Nonlinear Programming. Stanford University Press, Stanford
6. Aubin, J.-P. (1984): L'Analyse Non Linéaire et Ses Motivations Économiques. Masson, Paris
7. Aussel, D., Corvellec, J.-N., Lassonde, M. (1994): Subdifferential Characterization of Quasiconvexity and Convexity. Journal of Convex Analysis 1, 195–201
8. Avriel, M., Diewert, W.E., Schaible, S., Zang, I. (1988): Generalized Convexity. Plenum Press, New York
9. Baiocchi, C., Capelo, A. (1984): Variational and Quasivariational Inequalities. Applications to Free Boundary Problems. John Wiley and Sons, New York
10. Bakushinskii, A.B. (1979): Equivalent Transformations of Variational Inequalities with Applications. Soviet Mathematics Doklady 247, 1297–1300
11. Bakushinskii, A.B., Goncharskii, A.V. (1989): Iterative Solution Methods for Ill-Posed Problems. Nauka, Moscow (in Russian)
12. Bakushinskii, A.B., Polyak, B.T. (1974): On the Solution of Variational Inequalities. Soviet Mathematics Doklady 15, 1705–1710
13. Baushke, H.H., Borwein, J.M. (1996): On Projection Algorithms for Solving Convex Feasibility Problems. SIAM Review 38, 367–426
14. Belen'kii, V.Z., Volkonskii, V.A. (Eds.) (1974): Iterative Methods in Game Theory and Programming. Nauka, Moscow (in Russian)
15. Berge, C. (1957): Théorie Générale des Jeux a n Personnes. Gauthier-Villars, Paris
16. Bertsekas, D.P. (1982): Constrained Optimization and Lagrange Multiplier Methods. Academic Press, New York
17. Bianchi, M. (1993): Pseudo P-monotone Operators and Variational Inequalities. Research Report No.6, Istituto di Econometrica e Matematica per le Decisioni Economiche, Universita Cattolica del Sacro Cuore, Milan
18. Bianchi, M., Schaible, S. (1996): Generalized Monotone Bifunctions and Equilibrium Problems. Journal of Optimization Theory and Applications 90, 31–43

19. Billups, S.C., Ferris, M.C. (1997): QPCOMP: A Quadratic Programming Based Solver for Mixed Complementarity Problems. Mathematical Programming **76**, 533–562

20. Blum, E., Oettli, W. (1993): Variational Principles for Equilibrium Problems. In: Guddat, J., Jongen, H.Th., Kummer, B., Nožička, F. (Eds.): Parametric Optimization and Related Topics III. Peter Lang Verlag, Frankfurt am Main, 79–88

21. Blum, E., Oettli, W. (1994): From Optimization and Variational Inequalities to Equilibrium Problems. The Mathematics Student **63**, 127–149

22. Border, K.C. (1985): Fixed Point Theorems with Applications to Economics and Game Theory. Cambridge University Press, Cambridge

23. Brézis, H., Nirenberg, L., Stampacchia, G. (1972): A Remark on Ky Fan's Minimax Principle. Bolletino della Unione Matematica Italiana **6**, 293–300

24. Browder, F.E. (1965): Multivalued Monotone Nonlinear Mappings and Duality Mappings in Banach Spaces. Transactions of the American Mathematical Society **71**, 780–785

25. Browder, F.E. (1966): On the Unification of the Calculus of Variations and the Theory of Monotone Nonlinear Operators in Banach Spaces. Proceeding of the National Academy of Sciences, USA **56**, 419–425

26. Bruck, R. (1977): On Weak Convergence of an Ergodic Iteration for the Solution of Variational Inequalities for Monotone Operators in Hilbert Space. Journal of Mathematical Analysis and Applications **61**, 159–164

27. Bulavsky, V.A., Kalashnikov, V.V. (1994): One-Parametric Method for Determining the State of Equilibrium. Economika i Matematicheskie Metody **30**, 129–138 (in Russian)

28. Céa, J. (1971): Optimisation: Théorie et Algorithmes. Dunod, Paris

29. Chepurnoi, N.D. (1982): Relaxation Method of Minimization of Convex Functions. Doklady Akademii Nauk Ukrainy. Series A **3**, 68–69 (in Russian)

30. Cimmino, G. (1938): Calcolo Approsimato per le Soluzioni dei Sistemi de Equazioni Lineari. La Ricerca Scientifica XVI **1**, 326–333

31. Clarke, F.H. (1983): Optimization and Nonsmooth Analysis. John Wiley and Sons, New York

32. Cohen, G. (1988): Auxiliary Problem Principle Extended to Variational Inequalities. Journal of Optimization Theory and Applications **59**, 325–333

33. Cottle, R.W., Giannessi, F., Lions, J.-L. (Eds.) (1980): Variational Inequality and Complementarity Problems. John Wiley and Sons, New York

34. Cottle, R.W., Pang, J.-S., Stone, R.E. (1992): The Linear Complementarity Problem. Academic Press, New York

35. Cournot, A. (1838): Recherches sur les Principles Mathématiques de la Théorie des Richesses. Paris, 1838 (Engl. transl.: Researches into the Mathematical Principles of the Theory of Wealth. Bacon N. (Ed.): Macmillan, New York, 1897)

36. Dafermos, S. (1983): An Iterative Scheme for Variational Inequalities. Mathematical Programming **26**, 40–47

37. Dafermos, S. (1990): Exchange Price Equiilibria and Variational Inequalities. Mathematical Programming **46**, 391–402

38. Daniilidis, A., Hadjisavvas, N. (1995): Variational Inequalities with Quasimonotone Multivalued Operators. Working Paper, Department of Mathematics, University of the Aegean, Samos, Greece, March 1995

39. Daniilidis, A., Hadjisavvas, N. On the Subdifferentials of Quasiconvex and Pseudoconvex Functions and Cyclic Monotonicity. Journal of Mathematical Analysis and Applications, to appear

40. Daniilidis, A., Hadjisavvas, N. (1999): Characterization of Nonsmooth Semi-strictly Quasiconvex and Strictly Quasiconvex Functions. Journal of Optimization Theory and Applications 102, 525–536
41. Dem'yanov, V.F., Malozemov, V.N. (1972): Introduction to Minimax. Nauka, Moscow (Engl. transl. in John Wiley and Sons, New York,1974)
42. Dem'yanov, V.F., Rubinov, A.M. (1968): Approximate Methods for Solving Extremum Problems. Leningrad University Press, Leningrade (Engl. transl. in Elsevier Science B.V., Amsterdam, 1970)
43. Dem'yanov, V.F., Rubinov, A.M. (1990): Principles of Nonsmooth Analysis and Quasidifferential Calculus. Nauka, Moscow (in Russian)
44. Dem'yanov, V.F., Vasil'yev, L.V. (1981): Nondifferentiable Optimization. Nauka, Moscow (Engl. transl. in Optimization Software, New York,1985)
45. Duvaut, G., Lions, J.-L. (1972): Les Inéquations en Mechanique et Physique. Dunod, Paris
46. Dyubin, G.N., Suzdal', V.G. (1981): Introduction to Applied Game Theory. Nauka, Moscow (in Russian)
47. Eaves, B.C. (1971): On the Basic Theorem of Complementarity. Mathematical Programming 1, 68–75
48. Eaves, B.C. (1976): A Finite Algorithm for the Linear Exchange Model. Journal of Mathematical Economics 3, 197–204
49. Ekeland, I., Temam, R. (1979): Convex Analysis and Variational Problems. North - Holland, Amsterdam
50. Eremin, I.I. (1965): The Relaxation Method of Solving Systems of Inequalities with Convex Functions on the Left-hand Sides. Soviet Mathematics Doklady 6, 219–222
51. Eremin, I.I., Mazurov, V.D. (1979): Non-Stationary Processes of Mathematical Programming. Nauka, Moscow (in Russian)
52. Fan Ky (1972): A Minimax Inequality and Applications. In: Shisha, O. (Ed.): Inequalities. III Academic Press, New York, 103–113
53. Fang, S.C., Petersen, E.L. (1982): Generalized Variational Inequalities. Journal of Optimization Theory and Applications 38, 363–383
54. Ferris, M.C. (1991): Finite Termination of the Proximal Point Algorithm. Mathematical Programming 50, 359 – 366
55. Ferris, M.C., Pang, J.-S. (1997): Engineering and Economic Applications of Complementarity Problems. SIAM Review 39, 669–713
56. Fichera, G. (1964): Problemi Elastostatici con Vincoli Unilaterali; il Problema di Signorini con Ambigue al Contorno. Atti della Academia Nazionale dei Lincei. Memorie. Classe di Scienze Fisiche, Matematiche e Naturali. Sezione I 8, 91–140.
57. Fisher, A. (1997): Solution of Monotone Complementarity Problems with Locally Lipschitzian Functions. Mathematical Programming 76, 513–532
58. Frank, M., Wolfe, P. (1956): An Algorithm for Quadratic Programming. Naval Research Logistics Quarterly 3, 95–110
59. Fukushima, M. (1992): Equivalent Differentiable Optimization Problems and Descent Methods for Asymmetric Variational Inequality Problems. Mathematical Programming 53, 99–110
60. Fukushima, M. (1996): Merit Functions for Variational Inequality and Complementarity Problems. In: Di Pillo, N., Giannessi, F. (Eds.): Nonlinear Optimization and Applications. Plenum Press, New York, 155–170
61. Gafni, E.M., Bertsekas, D.P. (1984): Two–Metric Projection Methods for Constrained Optimization. SIAM Journal on Control and Optimization 22, 936–964
62. Gale, D. (1960): The Theory of Linear Economic Models. McGraw-Hill, New York

63. Gale, D. (1976): The Linear Exchange Model. Journal of Mathematical Economics **3**, 205–209

64. Gill, P.E., Murray, W., Wright, M.H. (1981): Practical Optimization. Academic Press, New York

65. Girsanov, I.V. (1970): Lectures in Mathematical Theory of Extremum Problems. Moscow University Press, Moscow (Engl. transl. in Springer-Verlag, Berlin, 1972)

66. Glowinski, R., Lions, J.-L., Trémolières, R. (1976): Analyse Numerique des Inéquations Variationnelles. Dunod, Paris

67. Gol'shtein, E.G., Nemirovskii, A.S., Nesterov, Yu.E. (1995): Level Method, Its Extensions and Applications. Ekonomika i Matematicheskie Metody **31**, 164–181 (in Russian)

68. Gol'shtein, E.G., Tret'yakov, N.V. (1989): Augmented Lagrange Functions. Nauka, Moscow (Engl. transl. in John Wiley and Sons, New York, 1996)

69. Hadjisavvas, N., Schaible, S. (1993): On Strong Pseudomonotonicity and (Semi) Strict Quasimonotonicity. Journal of Optimization Theory and Applications **79**, 139–155.

70. Harker, P.T., Pang, J.-S. (1990): Finite-Dimensional Variational Inequality and Nonlinear Complementarity Problems: A Survey of Theory, Algorithms and Applications. Mathematical Programming **48**, 161–220

71. Harker, P.T., Xiao, B. (1990): Newton's Method for the Nonlinear Complementarity Problem: A B-differentiable Equation Approach. Mathematical Programming **48**, 339–357

72. Hartmann, P., Stampacchia, G. (1966): On Some Nonlinear Elliptic Differential Functional Equations. Acta Mathematica **115**, 153–188

73. Hassouni, A. (1992): Quasimonotone Multifunctions; Applications to Optimality Conditions in Quasiconvex Programming. Numerical Functional Analysis and Optimization **13**, 267–275

74. He, B. (1992): A Projection and Contraction Method for a Class of Linear Complementarity Problems and Its Applications in Convex Quadratic Programming. Applied Mathematics and Optimization **25**, 247–262

75. He, B. (1994): Solving a Class of Linear Projection Equations. Numerische Mathematik **68**, 71–80

76. He, B. (1996): A Modified Projection and Contraction Method for a Class of Linear Complementarity Problems. Journal of Computational Mathematics **14**, 54–63

77. He, B. (1997): A Class of Projection and Contraction Methods for Monotone Variational Inequalities. Applied Mathematics and Optimization **35**, 69–76

78. Hiriart-Urruty, J.B., Lemaréchal, C. (1993): Convex Analysis and Minimization Algorithms. Springer-Verlag, Berlin (two volumes)

79. Hoffman, A.J. (1952): On Approximate Solutions of Systems of Linear Inequalities. Journal of Research at National Bureau of Standards **49**, 263–265

80. Isac, G. (1992): Complementarity Problems. Springer-Verlag, Berlin

81. Iusem, A.N. (1994): An Iterative Algorithm for the Variational Inequality Problem. Computational and Applied Mathematics **13**, 103–114

82. Josephy, N.H. (1979): Newton's Method for Generalized Equations. Technical Report No. 1965, Mathematics Research Center, University of Wisconsin, Madison

83. Kaczmarz, S. (1937): Angenäherte Auflösung von Systemen linearer Gleichungen. Bulletin Internationel de l'Académie Polonaise des Sciences et des Lettres. Classe des Sciences Mathématiques et Naturelles. Séries A **60**, 596–599

84. Kantorovich, L.V., Akilov, G.P. (1959): Functional Analysis in Normed Spaces. Fizmatgiz, Moscow (in Russian)

85. Karamardian, S. (1971): Generalized Complementarity Problems. Journal of Optimization Theory and Applications **8**, 161–167
86. Karamardian, S. (1976): An Existence Theorem for the Complementarity Problem. Journal of Optimization Theory and Applications **19**, 227–232
87. Karamardian, S. (1976): Complementarity over Cones with Monotone and Pseudomonotone Maps. Journal of Optimization Theory and Applications **18**, 445–454
88. Khobotov, E.N. (1987): Modification of the Extragradient Method for Solving Variational Inequalities and Certain Optimization Problems. USSR Computational Mathematics and Mathematical Physics **27**, 120–127
89. Kinderlerer, D., Stampacchia, G. (1980): An Introduction to Variational Inequalities and Their Applications. Academic Press, New York
90. Kiwiel, K.C. (1983): An Aggregate Subgradient Method for Nonsmooth Convex Minimization. Mathematical Programming **27**, 320–341
91. Kiwiel, K.C. (1985): Methods of Descent for Nondifferentiable Optimization. Springer-Verlag, Berlin
92. Kiwiel, K.C. (1995): Block-Iterative Surrogate Projection Methods for Convex Feasibility Problems. Linear Algebra and Its Applications **215**, 225–259
93. Kiwiel, K.C. (1995): Proximal Level Bandle Methods for Convex Nondifferentiale Optimization, Saddle Point Problems and Variational Inequalities. Mathematical Programming **69**, 89–109
94. Komlósi, S. (1994): Generalized Monotonicity in Nonsmooth Analysis, In: Komlósi, S., Rapcsák, T., Schaible, S. (Eds.): Generalized Convexity. Springer-Verlag, Heidelberg, 263–275
95. Konnov, I.V. (1982): A Subgradient Method of Successive Relaxation for Solving Optimization Problems. Preprint VINITI No. 531-83, Faculty of Computational Mathematics and Cybernetics, Kazan University, Kazan (in Russian)
96. Konnov, I.V. (1984): An Application of the Conjugate Subgradient Method for Minimization of Quasiconvex Functions. Issledovaniya po Prikladnoi Matematike **12**, 46–58 (Engl. transl. in Journal of Soviet Mathematics **45** (1989), 1019–1026)
97. Konnov, I.V. (1984): A Method of the Conjugate Subgradient Type for Minimization of Functionals. Issledovaniya po Prikladnoi Matematike **12**, 59–62 (Engl. transl. in Journal of Soviet Mathematics **45** (1989), 1026–1029).
98. Konnov, I.V. (1990): On Properties of Support and Quasi-Support Vectors. Issledovaniya po Prikladnoi Matematike **17**, 50–57 (in Russian)
99. Konnov, I.V. (1990): Convergence of Relaxation Methods for Solving Problems of Nondifferentiable Optimization with Constraints. Issledovaniya po Prikladnoi Matematike **17**, 57–71 (in Russian)
100. Konnov, I.V. (1992): Estimates of the Labor Cost of Successive Relaxation Methods. Issledovaniya po Prikladnoi Matematike **19**, 34–51 (in Russian)
101. Konnov, I.V. (1992): Combined Subgradient Methods for Finding Saddle Points. Russian Mathematics (Iz. VUZ) **36**, no.10, 30–33
102. Konnov, I.V. (1993): A Two-Level Subgradient Method for Finding Saddle Points of Convex-Concave Functions. Computational Mathematics and Mathematical Physics **33**, 453–459
103. Konnov, I.V. (1993): Combined Relaxation Methods for Finding Equilibrium Points and Solving Related Problems. Russian Mathematics (Iz. VUZ) **37**, no.2, 44–51
104. Konnov, I.V. (1993): A Combined Method for Variational Inequalities. News-Letter of the Mathematical Programming Association **4**, 64 (in Russian)
105. Konnov, I.V. (1993): Methods of Nondifferentiable Optimization. Kazan University Press, Kazan (in Russian)

106. Konnov, I.V. (1993): On Combined Relaxation Method's Convergence Rates. Russian Mathematics (Iz. VUZ) **37**, no.12, 89–92
107. Konnov, I.V. (1994): Applications of the Combined Relaxation Method to Finding Equilibrium Points of a Quasi-Convex-Concave Function. Russian Mathematics (Iz. VUZ) **38**, no.12, 66–71
108. Konnov, I.V. (1995): Combined Relaxation with Decomposition for Finding Equilibrium Points. Computational Mathematics and Mathematical Physics **35**, 281–286
109. Konnov, I.V. (1995): Combined Relaxation Methods for Solving Vector Equilibrium Problems. Russian Mathematics (Iz. VUZ) **39**, no.12, 51–59
110. Konnov, I.V. (1996): On Application of Projective Transformations in Combined Relaxation Methods. In: Fedotov, E.M. (Ed.): Theory of the Mesh Methods for Nonlinear Boundary Problems, Proceedings of Russia Workshop, Kazan Mathematics Foundation, Kazan, 63–65 (in Russian)
111. Konnov, I.V. (1996): A General Approach to Finding Stationary Points and the Solution of Related Problems. Computational Mathematics and Mathematical Physics **36**, 585–593
112. Konnov, I.V. (1996): Vector Variational Inequalities and Inverse Vector Optimization Problems. In: Uglov, V.A., Sidorov, S.V. (Eds.): Multiple Criteria and Game Problems under Uncertainty. The Fourth International Workshop. Russian Correspondence Institute of Textile and Light Industry, Moscow, 44
113. Konnov, I.V. (1997): Combined Relaxation Methods Having a Linear Rate of Convergence. News-Letter of the Mathematical Programming Association **7**, 135–137 (in Russian)
114. Konnov, I.V. (1997): A Combined Method for Smooth Equilibrium Problems with Nonlinear Constraints. Optimization Methods and Software **7**, 311–324
115. Konnov, I.V. (1997): A Class of Combined Iterative Methods for Solving Variational Inequalities. Journal of Optimization Theory and Applications **94**, 677–693
116. Konnov, I.V. (1997): Combined Relaxation Methods for Solving Equilibrium Problems and Variational Inequalities. D. Sc. Thesis, Kazan University, Kazan (in Russian)
117. Konnov, I.V. (1997): On Systems of Variational Inequalities. Russian Mathematics (Iz. VUZ), **41**, no.12, 77–86
118. Konnov, I.V. (1998): A Combined Relaxation Method for Variational Inequalities with Nonlinear Constraints. Mathematical Programming **80**, 239–252
119. Konnov, I.V. (1998): Accelerating the Convergence Rate of a Combined Relaxational Method. Computational Mathematics and Mathematical Physics **38**, 49–56
120. Konnov, I.V. (1998): On the Convergence of Combined Relaxation Methods for Variational Inequalities. Optimization Methods and Software **80**, 239–252
121. Konnov, I.V. (1998): On Quasimonotone Variational Inequalities. Journal of Optimization Theory and Applications **99**, 165–181
122. Konnov, I.V. (1998): An Inexact Combined Relaxation Method for Multivalued Inclusions. Russian Mathematics (Iz. VUZ) **42**, 55–59
123. Konnov, I.V. (1999): Combined Relaxation Method for Decomposable Variational Inequalities. Optimization Methods and Software **10** (1999), 711–728
124. Konnov, I.V. (1999): Implementable Feasible Quasi-nonexpansive Operators. Russian Mathematics (Iz. VUZ) **43**, no.5, 30–34
125. Konnov, I.V. (1999): A Combined Method for Variational Inequalities with Monotone Operators. Computational Mathematics and Mathematical Physics **39**, 1051–1056

126. Konnov, I.V., Schaible, S. (2000): Duality for Equilibrium Problems under Generalized Monotonicity. Journal of Optimization Theory and Applications **104**, 395–408

127. Konnov, I.V., Yao, J.C. (1997): On the Generalized Vector Variational Inequality Problem. Journal of Mathematical Analysis and Applications **206**, 42–58

128. Korpelevich, G.M. (1976): Extragradient Method for Finding Saddle Points and Other Problems. Matecon **12**, 747–756

129. Korpelevich, G.M. (1979): Relaxation Methods of Solving Matrix Games. Mathematical Methods of Solving Economical Problems **8**, 21–31 (in Russian)

130. Kose, T. (1956): Solutions of Saddle Value Problems by Differential Equations. Econometrica, **24**, 59–70

131. Kulikov, A.N., Fazylov, V.R. (1984): A Finite Solution Method for Systems of Convex Inequalities. Soviet Mathematics (Iz.VUZ) **28**, no.11, 75–80

132. Kulikov, A.N., Fazylov, V.R. (1984): A Finite Linearization Method for Systems of Convex Inequalities. Kibernetika, no. 4, 36–40

133. Lemaréchal, C., Nemirovskii, A., Nesterov, Y. (1995): New Variants of Bundle Methods. Mathematical Programming **69**, 111–147

134. Lemke, C.E. (1965): Bimatrix Equilibrium Points and Mathematical Programming. Management Science **11**, 681–689

135. Lescarret, C. (1965): Cas d'Addition des Applications Monotones Maximales dan un Espace de Hilbert. Comptes Rendus Hebdomadaires des Séances de l'Academie des Sciences (Paris) **261**, 1160–1163

136. Levitin, E.S., Polyak, B.T. (1966): Constrained Minimization Methods. USSR Computational Mathematics and Mathematical Physics. **6**, 1–50

137. Lions, J.-L., Stampacchia, G. (1967): Variational Inequalities. Communications on Pure and Applied Mathematics **20**, 493–519

138. Lions, P.L., Mercier, B. (1979): Splitting Algorithms for the Sum of Two Monotone Operators. SIAM Journal on Numerical Analysis **16**, 964–979

139. Luc, D.T. (1993): Characterizations of Quasiconvex Functions. Bulletin of the Australian Mathematical Society **48**, 393–406

140. Luo, Z.-Q., Tseng, P. (1992): On the Linear Convergence of Descent Methods for Convex Essentially Smooth Minimization. SIAM Journal on Control and Optimization **30**, 408–425

141. Luo, Z.-Q., Tseng, P. (1992): Error Bound and Convergence Analysis of Matrix Splitting Algorithms for the Affine Variational Inequality Problem. SIAM Journal on Optimization **2**, 43–54

142. Luo, Z.-Q., Tseng, P. (1993): Error Bounds and the Convergence Analysis of Feasible Descent Methods: A General Approach. Annals of Operations Research **46**, 157–178

143. Magnanti, T.L., Perakis, G. (1995): A Unifying Geometric Framework and Complexity Analysis for Variational Inequalities. Mathematical Programming **71**, 327–352

144. Manne, A.S. (Ed.)(1985): Economic Equilibrium: Model Formulation and Solution. Mathematical Programming Study **23**, North-Holland, Amsterdam

145. Marcotte, P., Zhu, D.L. (1995): Global Convergence of Descent Processes for Solving Non Strictly Monotone Variational Inequalities. Computational Optimization and Applications **4**, 127–138

146. Martinet, B. (1970): Regularization d'Inéquations Variationnelles par Approximations Successives. Revue Française d'Informatique et de Recherche Opérationelle **4**, 154–159

147. Martos, B. (1975): Nonlinear Programming. Theory and Methods. Akadémiai Kiado, Budapest

148. Mathiesen, L. (1987): An Algorithm Based on a Sequence of Linear Complementarity Problems Applied to a Walras' Equilibrium Model: An Example. Mathematical Programming **37**, 1–18

149. Minty, G. (1967): On the Generalization of a Direct Method of the Calculus of Variations. Bulletin of the American Mathematical Society **73**, 315–321

150. Moreau, J.-J. (1965): Proximité et Dualité dans un Espace Hilbertien. Bulletin de Societe Mathematique de France **93**, 273–299

151. Mosco, U. (1976): Implicit Variational Problems and Quasivariational Inequalities. In: Gossez, J.P. et al., (Eds.): Nonlinear Operators and Calculus of Variations. Lecture Notes in Mathematics **543** Springer-Verlag, Berlin, 83–156

152. Motzkin, T.S., Schoenberg, I.J. (1954): The Relaxation Method for Linear Inequalities. Canadian Journal of Mathematics **6**, 393–404.

153. Murphy, F.H., Sherali, H.D., Soyster, A.L. (1982): A Mathematical Programming Approach for Determining Oligopolistic Market Equilibrium. Mathematical Programming **24**, 92–106

154. Nagurney, A. (1993): Network Economics: A Variational Inequality Approach. Kluwer Academic Publishers, Dordrecht

155. Nemirovskii, A.S. (1981): Effective Iterative Methods for Solving Equations with Monotone Operators. Ekonomika i Matematicheskie Metody (Matecon) **17**, 344–359

156. Nemirovskii, A.S., Yudin, D.B. (1979): Problem Complexity and Method Efficiency in Optimization. Nauka, Moscow (Engl. transl. in John Wiley and Sons, New York,1983).

157. Nenakhov, E.I., Primak, M.E. (1985): On Investigations of Equilibrium States of an Economic System. Kibernetika, no. 6, 93–99 (in Russian)

158. Nesterov, Yu., Nemirovsky, A.S. (1994): Interior-Point Polynomial Algorithms in Convex Programming. SIAM Publishers, Philadelphia

159. Nikaido, H. (1968): Convex Structures and Economic Theory. Academic Press, New York

160. Nikaido, H., Isoda, K. (1955): Note on Noncooperative Convex Games. Pacific Journal of Mathematics **5**, 807–815

161. Nirenberg, L. (1974): Topics in Nonlinear Functional Analysis. New York University Press, New York

162. Noor, M.A. (1987): General Nonlinear Variational Inequalities. Journal of Mathematical Analysis and Applications **126**, 78–84

163. Noor, M.A. (1993): General Algorithm for Variational Inequalities. Mathematica Japonica **38**, 47–53

164. Opoitsev, V.I. (1986): Nonlinear System Statics. Nauka, Moscow (in Russian)

165. Ortega, J.M., Rheinboldt, W.C. (1970): Iterative Solution of Nonlinear Equations in Several Variables. Academic Press, New York

166. Panagiotopoulos, P.D. (1985): Inequality Problems in Mechanics and Their Applications. Birkhauser, Boston

167. Pang, J.-S. (1985): Asymmetric Variational Inequality Problems over Product Sets: Applications and Iterative Methods. Mathematical Programming **31**, 206–219

168. Pang, J.-S. (1987): A Posteriori Error Bounds for the Linearly Constrained Variational Inequality Problem. Mathematics of Operations Research **12**, 474–484

169. Pang, J.-S. (1997): Error Bounds in Mathematical Programming. Mathematical Programming **79**, 299–332

170. Pang, J.-S., Chan, D. (1982): Iterative Methods for Variational and Complementarity Problems. Mathematical Programming **24**, 284–313

171. Passty, G.B. (1979): Ergodic Convergence to Zero of the Sum of Two Monotone Operators in Hilbert Space. Journal of Mathematical Analysis and Applications **72**, 383–390
172. Patriksson, M. (1997): Merit Functions and Descent Algorithms for a Class of Variational Inequality Problems. Optimization **41**, 37–55
173. Patriksson, M. (1999): Nonlinear Programming and Variational Inequality Problems: A Unified Approach. Kluwer Academic Publishers, Dordrecht
174. Penot, J.-P., Quang, P.H. (1997): Generalized Convexity of Functions and Generalized Monotonicity of Set-valued Maps. Journal of Optimization Theory and Applications **92**, 343–356
175. Polak, E. (1971): Computational Methods in Optimization: A Unified Approach. Academic Press, New York
176. Polak, E., Mayne, D.Q., Wardi, Y. (1983): On the Extension of Constrained Optimization Algorithms from Differentiable to Nondifferentiable Problems. SIAM Journal on Control and Optimization **21**, 179–203
177. Polterovich, V.M. (1990): Equilibrium and Economic Mechanism. Nauka, Moscow (in Russian)
178. Polyak, B.T. (1969): Minimization of Unsmooth Functionals. USSR Computational Mathematics and Mathematical Physics **9**, 14–29
179. Polyak, B.T. (1983): Introduction to Optimization. Nauka, Moscow, 1983 (Engl. transl. in Optimization Software, New York, 1987)
180. Popov, L.D. (1980): A Modification of the Arrow-Hurwicz Method of Finding Saddle Points. Matematicheskie Zametki **28**, 777–784 (in Russian)
181. Primak, M.E. (1984): An Algorithm for Finding a Solution of the Linear Pure Trade Market Model and the Linear Arrow-Debreu Model. Kibernetika, no. 5, 76–81 (in Russian)
182. Pshenichnyi, B.N. (1971): Necessary Conditions for an Extremum. Nauka, Moscow, 1969; (Engl. transl. in Marcel Dekker, New York, 1971)
183. Pshenichnyi, B.N., Danilin, Yu.M. (1978): Numerical Methods in Extremal Problems. MIR, Moscow
184. Robinson, S.M. (1981): Some Continuity Properties of Polyhedral Multifunctions. Mathematical Programming Study **17**, 206–214
185. Rockafellar, R.T. (1970): Convex Analysis. Princeton University Press, Princeton
186. Rockafellar, R.T. (1976): Monotone Operators and the Proximal Point Algorithm. SIAM Journal on Control and Optimization **14**, 877–898
187. Rockafellar, R.T. (1976): Augmented Lagrangians and the Proximal Point Algorithm in Convex Programming. Mathematics of Operations Research **1**, 97–116
188. Rockafellar, R.T. (1976): Monotone Operators and Augmented Lagrangian Methods in Nonlinear Programming. In: Mangasarian, O.L., Meyer, R.R., Robinson, S.M. (Eds.): Nonlinear Programming 3. Academic Press, New York, 1–25
189. Rockafellar, R.T. (1981): The Theory of Subgradients and Its Applications to Problems of Optimization. Heldermann-Verlag, Berlin
190. Rosen, J.B. (1960): The Gradient Projection Method for Nonlinear Programming I. Journal of the Society of Industrial and Applied Mathematics **8**, 181–217
191. Rosen, J.B. (1965): Existence and Uniqueness of Equilibrium Points for Concave n-Person Games. Econometrica **33**, 520–534
192. Saigal, R. (1976): Extension of the Generalized Complementarity Problem. Mathematics of Operations Research **1**, 260–266
193. Scarf, H.E., Hansen, T. (1973): Computation of Economic Equilibria. Yale Univ. Press, New Haven

194. Schaible, S. (1992): Generalized Monotone Maps. In: Giannessi, F. (Ed.): Proceedings of a Conference Held at "G. Stampacchia International School of Mathematics". Gordon and Breach Science Publishers, Amsterdam, 392–408
195. Schaible, S., Karamardian, S., Crouzeix, J.-P. (1993): Characterizations of Generalized Monotone Maps. Journal of Optimization Theory and Applications 76, 399–413
196. Schott, D. (1991): A General Iterative Scheme with Applications to Convex Optimization and Related Fields. Optimization 22, 885–902
197. Schrijver, A. (1986): Theory of Linear and Integer Programming. John Wiley and Sons, New York
198. Shih, M.H., Tan, K.K. (1988): Browder-Hartmann-Stampacchia Variational Inequalities for Multivalued Monotone Operators. Journal of Mathematical Analysis and Applications 134, 431–440
199. Shor, N.Z. (1977): Cut-off Method with Space Dilation in Convex Programming Problems. Cybernetics 13, 94–96
200. Shor, N.Z. (1977): New Development Trends in Nonsmooth Optimization Methods. Cybernetics 6, 87–91
201. Shor, N.Z. (1985): Minimization Methods for Non-differentiable Functions. Naukova Dumka, Kiev, 1979 (Engl. transl. in Springer-Verlag, Berlin, 1985)
202. Sibony, M. (1970): Méthodes Itératives pur les Équations et Inéquations aux Dérivées Partielles non Lin'eares de Type Monotone. Calcolo 7, 65–183
203. Solodov, M.V., Tseng, P. (1996): Modified Projection-type Methods for Monotone Variational Inequalities. SIAM Journal on Control and Optimization 34, 1814–1830
204. Stampacchia, G. (1964): Formes Bilinéaires Coercitives sur le Ensembles Convexes. Comptes Rendus Hebdomadaires des Séances de l'Academie des Sciences (Paris) 258, 4413–4416
205. Sun, D. (1994): A Projection and Contraction Method for the Nonlinear Complementarity Problem and Its Extensions. Chinese Journal of Numerical Mathematics and Applications 16, 73–84
206. Sun, D. (1995): A New Step-size Skill for Solving a Class of Nonlinear Projection Equations. Journal of Computational Mathematics 13, 357–368
207. Sun, D. (1996): A Class of Iterative Methods for Solving Nonlinear Projection Equations. Journal of Optimization Theory and Applications 91, 123–140
208. Szidarovsky, F., Yakowitz, S. (1977): A New Proof of the Existence and Uniqueness of the Cournot Equilibrium. International Economic Review 18, 787–789
209. Takayama, T., Judge, G.G. (1971): Spatial and Temporal Price and Allocation Models. North-Holland, Amsterdam
210. Todd, M.J. (1976): The Computations of Fixed Points and Applications. Springer-Verlag, Berlin
211. Tseng, P. (1995): On Linear Convergence of Iterative Methods for the Variational Inequality Problem. Journal of Computational and Applied Mathematics 60, 237–252
212. Uryas'ev, S.P. (1990): Adaptive Algorithms of Stochastic Optimization and Game Theory. Nauka, Moscow (in Russian)
213. Vainberg, M.M. (1973): Variational Methods and Method of Monotone Operators in the Theory of Nonlinear Equations. Nauka, Moscow, 1972 (Engl. transl. in John Wiley and Sons, New York, 1973)
214. Vasil'yev, F.P. (1981): Methods of Solving Extremal Problems. Nauka, Moscow (in Russian)
215. Vasin, V.V., Ageev, A.L. (1993): Ill-Posed Problems with A Priori Information. Nauka, Ekaterinburg, 1993 (Engl. transl. in VSP, Utrecht, 1995)

216. Walras, L. (1874): Eléments d'Économie Politique Pure. Corbaz, Lausanne
217. Wolfe, P. (1975): A Method of Conjugate Subgradients for Minimizing Non-differentiable Functions. In: Balinski, M.L., Wolfe, P. (Eds.): Nondifferentiable Optimization. Mathematical Programming Study **3**, 145–173
218. Yamashita, N., Fukushima, M. (1995): On Stationary Points of the Implicit Lagrangian for Nonlinear Complementarity Problems. Journal of Optimization Theory and Applications **84**, 653–663
219. Yao, J.C. (1994): Multi-valued Variational Inequalities with K - Pseudomonotone Operators. Journal of Optimization Theory and Applications **83**, 391–403
220. Yevtushenko, Yu.G. (1982): Methods of Solving Extremal Problems and Their Applications in Optimization Systems. Nauka, Moscow, 1982 (Engl. transl. in Optimization Software, New York, 1985)
221. Yudin, D.B., Nemirovskii, A.S. (1977): Informational Complexity and Efficient Methods for the Solution of Convex Extremal Problems, I, II. Matecon **13**, 25–45; 550–559
222. Zangwill, W.I. (1969): Nonlinear Programming: A Unified Approach. Prentice-Hall, Englewood Cliffs
223. Zangwill, W.I., Garcia, C.B. (1981): Equilibrium Programming: The Path-Following Approach and Dynamics. Mathematical Programming **21**, 262–289
224. Zhu, D.L., Marcotte, P. (1994): An Extended Descent Framework for Variational Inequalities. Journal of Optimization Theory and Applications **80**, 349–366
225. Zoutendijk, G. (1960): Methods of Feasible Directions. Elsevier Science B.V., Amsterdam
226. Zukhovitskii, S.I., Polyak, R.A., Primak, M.E. (1969): Two Methods of Search for Equilibrium Points of n-person Concave Games. Soviet Mathematics Doklady **10**, 279–282
227. Zukhovitskii, S.I., Polyak, R.A., Primak, M.E. (1970): On an n-person Concave Game and a Production Model. Soviet Mathematics Doklady **11**, 522 – 526

Index

Vol. 449: F. Fang, M. Sanglier (Eds.), Complexity and Self-Organization in Social and Economic Systems. IX, 317 pages. 1997.

Vol. 450: P. M. Pardalos, D. W. Hearn, W. W. Hager, (Eds.), Network Optimization. VIII, 485 pages, 1997.

Vol. 451: M. Salge, Rational Bubbles. Theoretical Basis, Economic Relevance, and Empirical Evidence with a Special Emphasis on the German Stock Market.IX, 265 pages. 1997.

Vol. 452: P. Gritzmann, R. Horst, E. Sachs, R. Tichatschke (Eds.), Recent Advances in Optimization. VIII, 379 pages. 1997.

Vol. 453: A. S. Tangian, J. Gruber (Eds.), Constructing Scalar-Valued Objective Functions. VIII, 298 pages. 1997.

Vol. 454: H.-M. Krolzig, Markov-Switching Vector Autoregressions. XIV, 358 pages. 1997.

Vol. 455: R. Caballero, F. Ruiz, R. E. Steuer (Eds.), Advances in Multiple Objective and Goal Programming. VIII, 391 pages. 1997.

Vol. 456: R. Conte, R. Hegselmann, P. Terna (Eds.), Simulating Social Phenomena. VIII, 536 pages. 1997.

Vol. 457: C. Hsu, Volume and the Nonlinear Dynamics of Stock Returns. VIII, 133 pages. 1998.

Vol. 458: K. Marti, P. Kall (Eds.), Stochastic Programming Methods and Technical Applications. X, 437 pages. 1998.

Vol. 459: H. K. Ryu, D. J. Slottje, Measuring Trends in U.S. Income Inequality. XI, 195 pages. 1998.

Vol. 460: B. Fleischmann, J. A. E. E. van Nunen, M. G. Speranza, P. Stähly, Advances in Distribution Logistic. XI, 535 pages. 1998.

Vol. 461: U. Schmidt, Axiomatic Utility Theory under Risk. XV, 201 pages. 1998.

Vol. 462: L. von Auer, Dynamic Preferences, Choice Mechanisms, and Welfare. XII, 226 pages. 1998.

Vol. 463: G. Abraham-Frois (Ed.), Non-Linear Dynamics and Endogenous Cycles. VI, 204 pages. 1998.

Vol. 464: A. Aulin, The Impact of Science on Economic Growth and its Cycles. IX, 204 pages. 1998.

Vol. 465: T. J. Stewart, R. C. van den Honert (Eds.), Trends in Multicriteria Decision Making. X, 448 pages. 1998.

Vol. 466: A. Sadrieh, The Alternating Double Auction Market. VII, 350 pages. 1998.

Vol. 467: H. Hennig-Schmidt, Bargaining in a Video Experiment. Determinants of Boundedly Rational Behavior. XII, 221 pages. 1999.

Vol. 468: A. Ziegler, A Game Theory Analysis of Options. XIV, 145 pages. 1999.

Vol. 469: M. P. Vogel, Environmental Kuznets Curves. XIII, 197 pages. 1999.

Vol. 470: M. Ammann, Pricing Derivative Credit Risk. XII, 228 pages. 1999.

Vol. 471: N. H. M. Wilson (Ed.), Computer-Aided Transit Scheduling. XI, 444 pages. 1999.

Vol. 472: J.-R. Tyran, Money Illusion and Strategic Complementarity as Causes of Monetary Non-Neutrality. X, 228 pages. 1999.

Vol. 473: S. Helber, Performance Analysis of Flow Lines with Non-Linear Flow of Material. IX, 280 pages. 1999.

Vol. 474: U. Schwalbe, The Core of Economies with Asymmetric Information. IX, 141 pages. 1999.

Vol. 475: L. Kaas, Dynamic Macroeconomics with Imperfect Competition. XI, 155 pages. 1999.

Vol. 476: R. Demel, Fiscal Policy, Public Debt and the Term Structure of Interest Rates. X, 279 pages. 1999.

Vol. 477: M. Théra, R. Tichatschke (Eds.), Ill-posed Variational Problems and Regularization Techniques. VIII, 274 pages. 1999.

Vol. 478: S. Hartmann, Project Scheduling under Limited Resources. XII, 221 pages. 1999.

Vol. 479: L. v. Thadden, Money, Inflation, and Capital Formation. IX, 192 pages. 1999.

Vol. 480: M. Grazia Speranza, P. Stähly (Eds.), New Trends in Distribution Logistics. X, 336 pages. 1999.

Vol. 481: V. H. Nguyen, J. J. Strodiot, P. Tossings (Eds.). Optimation. IX, 498 pages. 2000.

Vol. 482: W. B. Zhang, A Theory of International Trade. XI, 192 pages. 2000.

Vol. 483: M. Königstein, Equity, Efficiency and Evolutionary Stability in Bargaining Games with Joint Production. XII, 197 pages. 2000.

Vol. 484: D. D. Gatti, M. Gallegati, A. Kirman, Interaction and Market Structure. VI, 298 pages. 2000.

Vol. 485: A. Garnaev, Search Games and Other Applications of Game Theory. VIII, 145 pages. 2000.

Vol. 486: M. Neugart, Nonlinear Labor Market Dynamics. X, 175 pages. 2000.

Vol. 487: Y. Y. Haimes, R. E. Steuer (Eds.), Research and Practice in Multiple Criteria Decision Making. XVII, 553 pages. 2000.

Vol. 488: B. Schmolck, Ommitted Variable Tests and Dynamic Specification. X, 144 pages. 2000.

Vol. 489: T. Steger, Transitional Dynamics and Economic Growth in Developing Countries. VIII, 151 pages. 2000.

Vol. 490: S. Minner, Strategic Safety Stocks in Supply Chains. XI, 214 pages. 2000.

Vol. 491: M. Ehrgott, Multicriteria Optimization. VIII, 242 pages. 2000.

Vol. 492: T. Phan Huy, Constraint Propagation in Flexible Manufacturing. IX, 258 pages. 2000.

Vol. 493: J. Zhu, Modular Pricing of Options. X, 170 pages. 2000.

Vol. 494: D. Franzen, Design of Master Agreements for OTC Derivatives. VIII, 175 pages. 2001.

Vol. 495: I Konnov, Combined Relaxation Methods for Variational Inequalities. XI, 181 pages. 2001.